Thinking outwards 1

Thomas Nelson and Sons Ltd
36 Park Street London W1Y 4DE
PO Box 18123 Nairobi Kenya
Thomas Nelson (Australia) Ltd
597 Little Collins Street Melbourne 3000
Thomas Nelson and Sons (Canada) Ltd
81 Curlew Drive Don Mills Ontario
Thomas Nelson (Nigeria) Ltd
PO Box 336 Apapa Lagos
Thomas Nelson and Sons (South Africa) (Proprietary) Ltd
51 Commissioner Street Johannesburg

First published 1971

0 17 421039 6

Filmset and Printed Offset Litho
by Cox and Wyman Limited
London, Fakenham and Reading

Understanding new mathematics

George W. Rodda B.Sc.

Principal Lecturer in Mathematics,
Alsager College of Education, Cheshire

Thinking outwards 1

Nelson

Contents

Let's Discover **Squares and Rectangles**

1. Look around your classroom and your school. See how many floors, walls and ceilings you can find which are tiled with **squares**.

A wall might look like this:

A
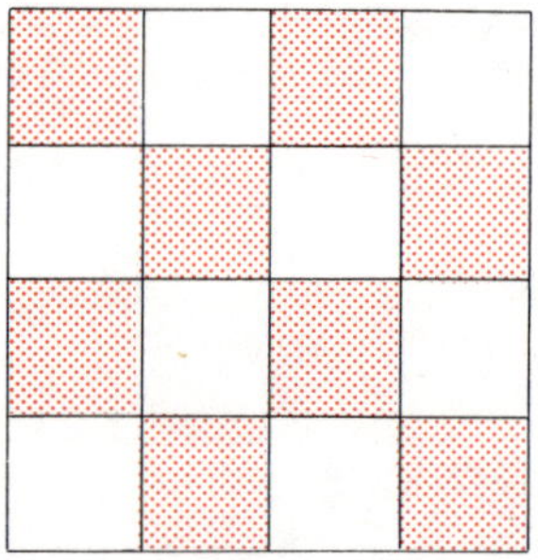

A floor might look like this:

B
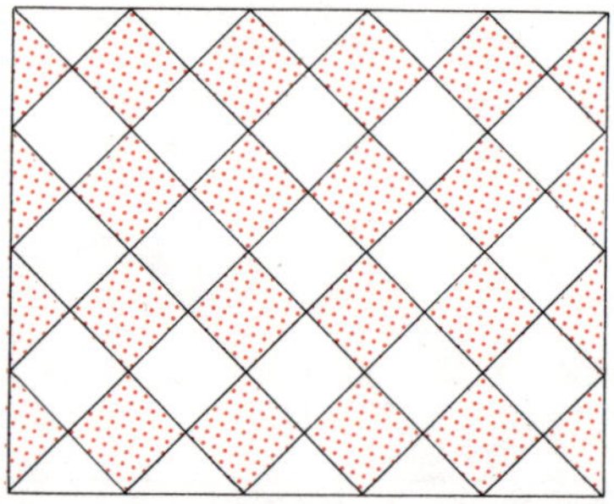

2. Are the two patterns the same? Look at A carefully. Now turn your book and look at B.
The patterns A and B are really the same. They are both made of squares which are fitted together.

- When plane shapes are fitted together the shapes are being tessellated.

Here are two different tessellations with 1-centimetre squares.

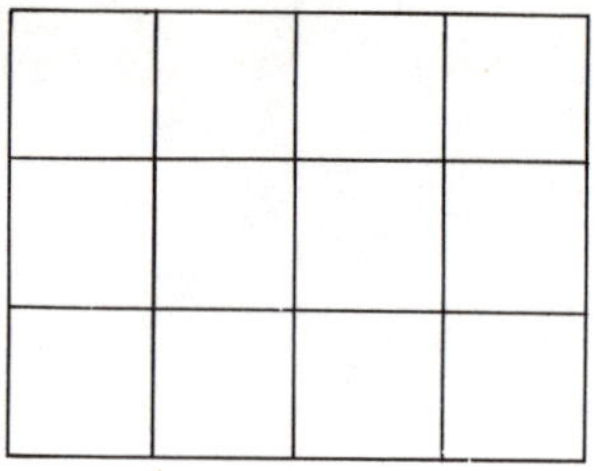

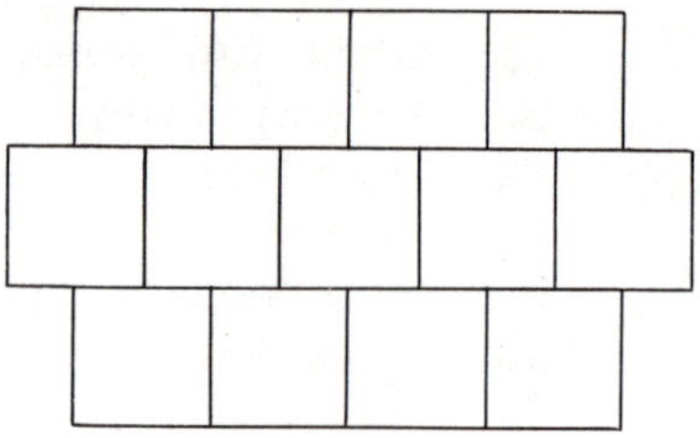

Circular tiles are not used because circles cannot be tessellated.

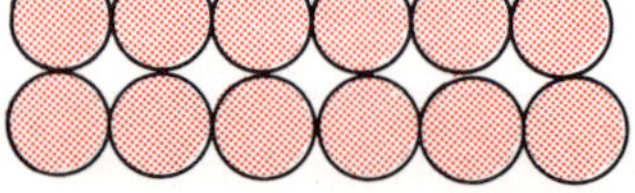

Rectangles can be tessellated.

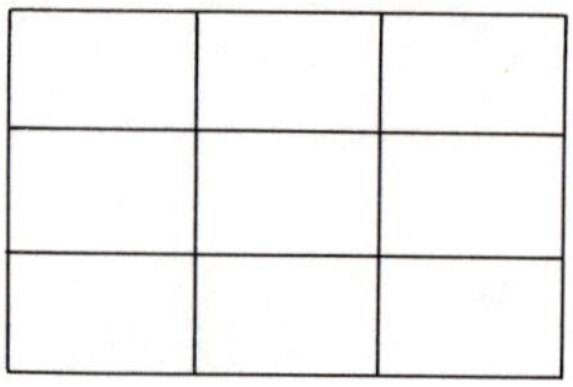

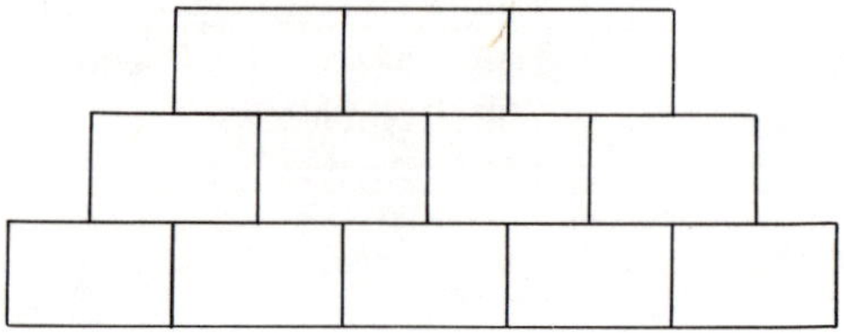

3. A brick wall is probably the most common example of rectangles being tessellated. Try to find or draw some more examples.

Here are four squares D and four rectangles E being brought together.

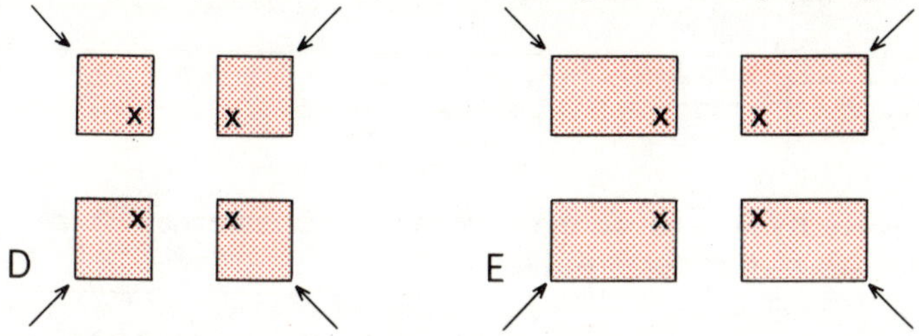

In D and in E the four angles marked x must fit snugly together.

4. Cut out a square, tear off the four corners and fit them together.
5. Cut out a rectangle, tear off the four corners and fit them together.
6. Try tessellating using squares and rectangles together.

Let's Try

1. PQRS is a rectangle.
   a. How many rectangles can you find in this drawing?
   b. How many squares can you find in this drawing?

2. Cut out a piece of card which is 3 cm long and 2 cm wide. What is the smallest number of tiles like your piece of card that you would need to make a square?

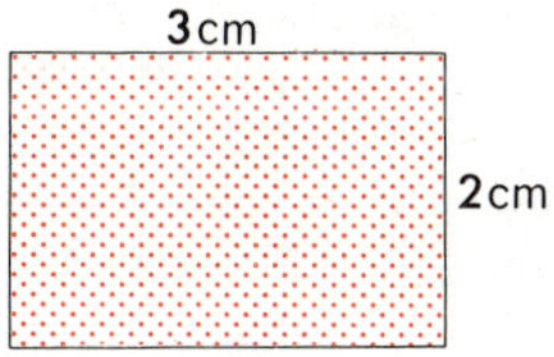

3. This shape is called a hexominoe because it is made up of six $\frac{1}{2}$-cm squares. Cut out a shape like this and use it to build up a tessellation.

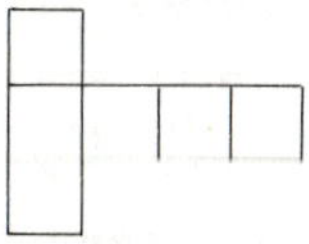

4. Here is another hexominoe. Cut out a shape like this and use it to build up a tessellation.

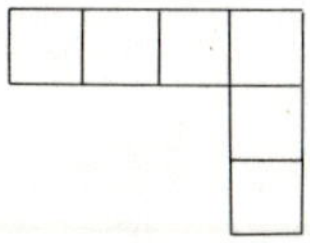

5. Copy and cut out this shape twice and cut along AB on both pieces. Try to arrange your four pieces into a square.

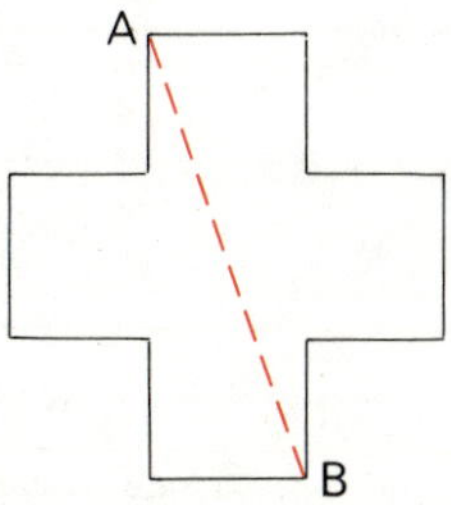

Let's Discover — Factors and Trees

12÷4 = 3 rem. 0	12÷5 = 2 rem. 2

Remember that **12 is divisible by 4** because the remainder is 0, and that **12 is not divisible by 5** because the remainder is not 0.

1. Complete the following:

a. 4×3 = . . . **b.** . . . ×2 = 12 **c.** 1× . . . = 12

2. Is 12 divisible by: **a.** 1 **b.** 2 **c.** 3 **d.** 4 **e.** 6 **f.** 12

{1, 2, 3, 4, 6, 12} **is the set of factors of 12.**

3. Complete this set of factors of **a.** 16 {1, 2, . . . , 8, 16}
b. 24 {1, 2, . . . , . . . , . . . , 12, 24}

4. Do you know what a family tree looks like? Ask your teacher if you don't.

5. a. Here is a **factor tree** for 24. Start at the top 24 and follow the branches.

b. Look at the bottom line of the tree. Is it true that 24 = 2×2×2×3?

c. We can pick out some factors of 24 from this tree. Each number which is a factor has been shown in orange. Write down the set of factors of 24 from this tree. Which factors of 24 are missing from your set?

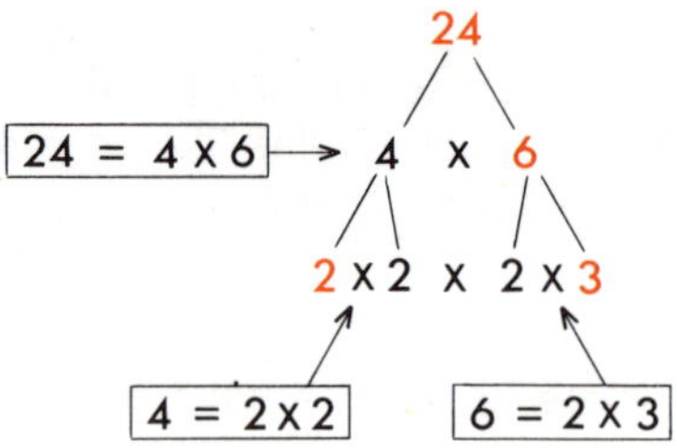

6. a. Complete these factor trees for 24:

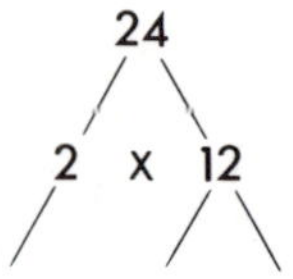

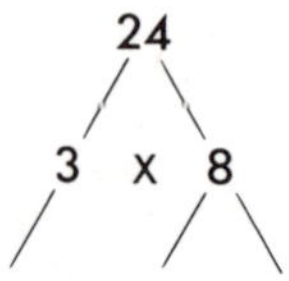

b. Were your last rows the same in both trees?

c. Is 2×2×2×3 = 3×2×2×2?

7. Jane wished to multiply a number by 14.
Her friend said multiply the number by 2, and then your product by 7.

a. Here is part of Jane's working.
Finish the working for her.

$$27 \times 14 = \begin{array}{r} 27 \\ \times\ 2 \\ \hline 54 \\ \times\ 7 \\ \hline \\ \hline \end{array}$$

b. Are 2 and 7 factors of 14?

c. Is $2 \times 7 = 14$?

- This is the **factor method of multiplying**.

Let's Try

1. Find the replacement for each $\triangle$ in:

a. $2 \times \triangle = 10$	b. $\triangle \times 5 = 15$	c. $4 \times \triangle = 12$	d. $\triangle \times 6 = 24$
e. $6 \times \triangle = 42$	f. $\triangle \times 8 = 32$	g. $9 \times \triangle = 36$	h. $\triangle \times 5 = 50$

2. Complete these factor trees for 16 and 36:

a.
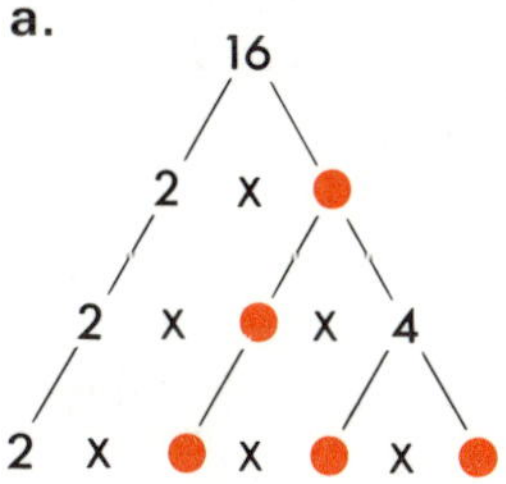

b.
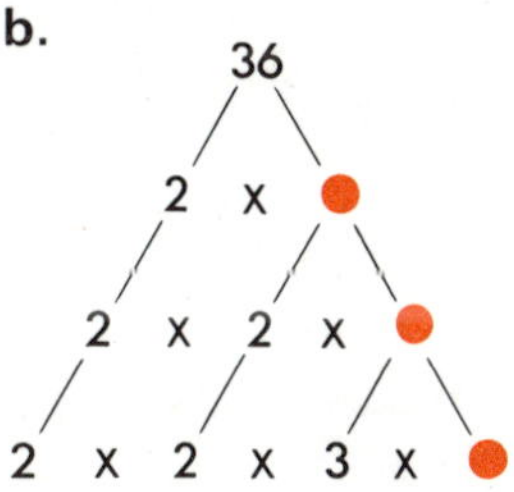

c. Write the set of all factors of 16.

d. Write the set of all factors of 36.

3. Without drawing trees list the set of factors of:

a. 15 b. 20 c. 18 d. 45 e. 33 f. 23

4. Draw factor trees for:

a. 18 b. 25 c. 27 d. 32 e. 45 f. 40

5. Use the factor method of multiplying to find:

a. 76×14 b. 104×15 c. 142×16 d. 91×20 e. 114×24

6. Tom said, "With the exception of 1, 9 is the smallest factor of 81." Bill said, "9 can't be the smallest factor of 81 because $3 \times 3 = 9$."
Who is right, Tom or Bill?

Let's Discover **Some Products**

Many years ago a clever method of multiplying one number by another was discovered by Russian peasants.

Look carefully at this example of the way they multiplied.

To multiply 25 by 11, divide the left-hand number by 2. Ignore remainders.

25×11		
12×22	(11×2)	Double the right-hand number.
6×44	(22×2)	
3×88	(44×2)	
1×176	(88×2)	

The peasants didn't like lines of working which started with an even number. They crossed out these lines and then totalled the rest of the right-hand numbers.

25×11	11
~~12×22~~	
~~6×44~~	
3×88	88 +
1×176	176
	275 = 25×11

1. Check the answer to 25×11 by the **factor method**.
Remember that $25 \times 11 = 11 \times 25$.

2. Using your multiplication tables you can find products in this way.
Fill in the missing numbers in:

a. $16 \times 14 = 16 \times (10+4)$
$= (16 \times 10) + (16 \times 4)$
$= \ldots + \ldots$
$= 224$

b. $34 \times 21 = 34 \times (20+1)$
$= (34 \times 20) + (34 \times 1)$
$= \ldots + \ldots$
$= 714$

The products in question **2** could have been found by setting out like this:

	16
	×14
$16 \times 10 \rightarrow$	160
$16 \times 4 \rightarrow$	64
$16 \times 14 \rightarrow$	224

	34
	×21
$34 \times 20 \rightarrow$	680
$34 \times 1 \rightarrow$	34
$34 \times 21 \rightarrow$	714

3. Fill in the missing numbers in order to check the products.

a.

	22
	×14
$22 \times 10 \rightarrow$	
$22 \times 4 \rightarrow$	
$22 \times 14 \rightarrow$	308

b.

	23
	×22
$23 \times 20 \rightarrow$	
$23 \times 2 \rightarrow$	
$23 \times 22 \rightarrow$	506

Let's Try

1. Use the Russian Peasant method to find the products of:

   a. 19×11 b. 24×13 c. 26×9 d. 47×22

2. Complete these multiplications:

   a. $14 \times 12 = 14 \times (10+2)$
   $= \ldots$

   b. $23 \times 13 = 23 \times (10+3)$
   $= \ldots$

   c. $23 \times 11 = 23 \times (10+1)$
   $= \ldots$

   d. $42 \times 21 = 42 \times (20+1)$
   $= \ldots$

3. Use the method shown at the bottom of the opposite page to find the products of:

   a. 19×14 b. 26×24 c. 29×23

   d. 32×64 e. 48×47 f. 47×74

4. Track for a model railway can be bought in short lengths of 20 cm and long lengths of 35 cm.

   a. What is the total length of 35 short lengths?

   b. What is the total length of 20 long lengths?

   c. A boy has 8 short lengths and 15 long lengths.
   What is his total length of track?

5. At 4p per kilogramme how much will a farmer receive for 135 kg of potatoes?

6. The top speed of an aeroplane is 76 kilometres per hour. How far can it travel in:

   a. 12 hours   b. 24 hours   c. 18 hours

7. In order to buy a bicycle on hire purchase I paid 75p per week for a year.
   How much did the bicycle cost?

8. I am going to lay 16 paving stones end to end. If each stone is 75 cm long how long will the path be in:

   a. centimetres   b. metres

9. From a roll of ribbon I can cut 16 lengths of 25 cm and have 7 cm left.
   How long is the roll of ribbon?

10. The sum of two numbers is 32. One number is 2 more than the other.
    What is the product of the two numbers?

Let's Discover

Series

John always seemed to be breaking windows, so his father said, "I will charge you 1p for the first, 2p for the second and so on. For each window, you will have to pay the same number of pence as the number of windows you have broken."

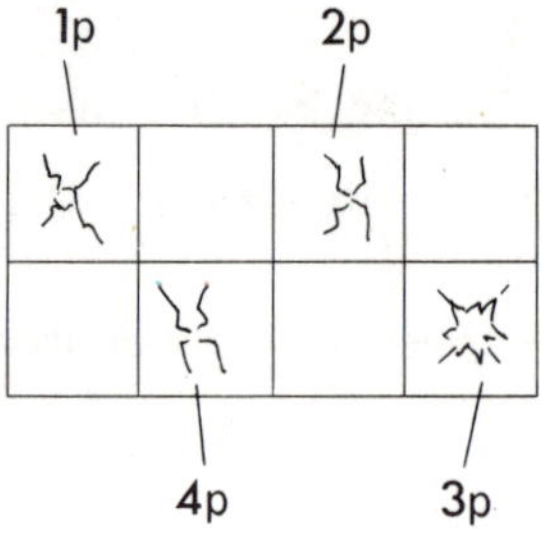

1. Complete the following:

a. For 3 windows it will cost John 1+2+3 = . . . p

b. For 5 windows it will cost John 1+2+3+4+5 = . . . p

Instead of going up in steps of one penny like this: 1, 2, 3, 4, 5, . . .
John's father decided that steps of 2p would be better like this: 1, 3, 5, 7, 9, 11, . . .
so John's **third** window would cost 5p.

2. Now how much does it cost for:

a. 3 windows **b.** 5 windows **c.** 6 windows

• 1, 3, 5, 7, 9, 11, . . . is called a **series of numbers.**

We can look for the pattern in a series of numbers. The next example looks at the pattern in the series 2, 4, 6, 8, 10, . . .

3. Fill in the missing numbers.

a. The next two numbers in the series 2, 4, 6, 8, 10, . . . are ____ and ____.

b. The series is increasing in steps of ____.

You can guess the next two numbers in these two series.

(A) 20, 18, 16, 14, 12, . . . (B) 1, 4, 7, 10, . . .

4. **a.** Which of the series (A) and (B) is increasing in steps of three (+3)?

b. Which of the series (A) and (B) is decreasing in steps of two (−2)?

5. **Not all series change in even steps.** Here is a series which has a pattern: 2, 6, 4, 8, 6, 10, 8, 12, . . .

a. Trace the series up to the 12 on a number line like this.

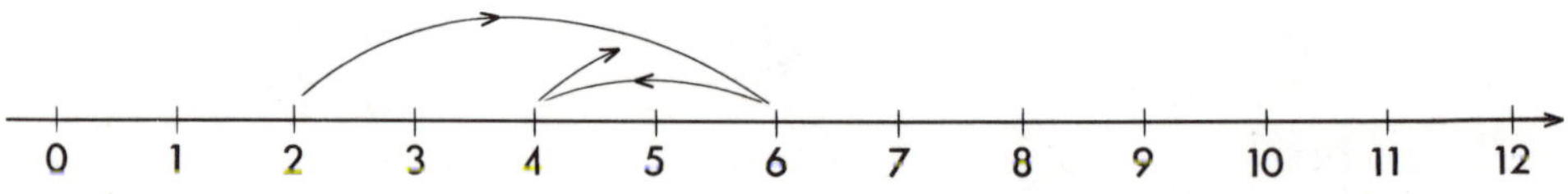

b. Complete the following:
We add . . . to get the second number. We subtract . . . to get the third number.

c. Write the next two numbers of the series
2, 6, 4, 8, 6, 10, 8, 12, . . .

- **Tracing the path of a series on a number line can often help you discover the pattern in the series.**

6. **a.** Here is a series of numbers which you can trace on a number line.
1, 3, 6, 10, 15, . . .

b. Fill in the empty boxes for the series, 1, 3, 6, 10, . . .

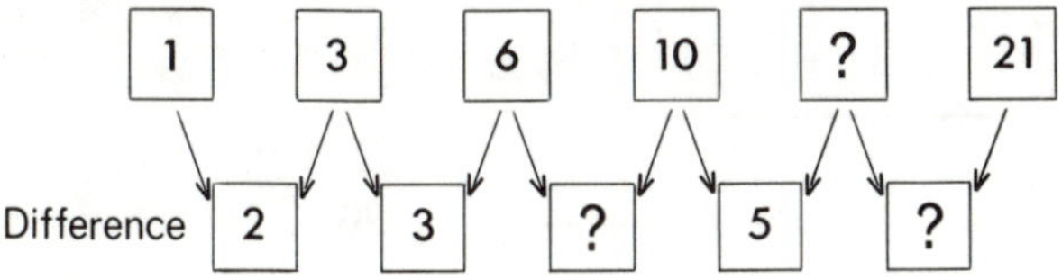

Let's Try

1. On a 100 square:

a. Put a line / through each number in the series 2, 4, 6, 8, 10, . . .

b. Put a line \ through each number in the series 3, 6, 9, 12, . . .

c. Write down the series of numbers which have X on them.
Check the pattern in the series.

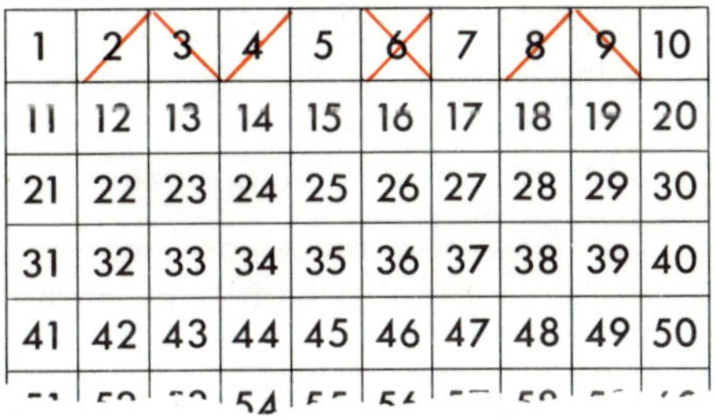

2. Make up some more questions like number **1**. For example, use series increasing in twos and fives.

3. Draw a number line and trace each of these series.
Write down the next two numbers in each series.

a. 1, 2, 4, 7, 11, . . . **b.** 0, 3, 2, 5, 4, . . .
c. 12, 8, 11, 7, 10, . . . **d.** 1, 10, 2, 9, 3, . . .

4. Find out all you can about:

a. Pascal's triangle **b.** Fibonacci series

Let's Discover — Square Numbers

On a geo-board (or squared paper) make up some squares. For example, on my squared paper I have shaded two squares, one containing 1 small square and one containing 4 small squares.

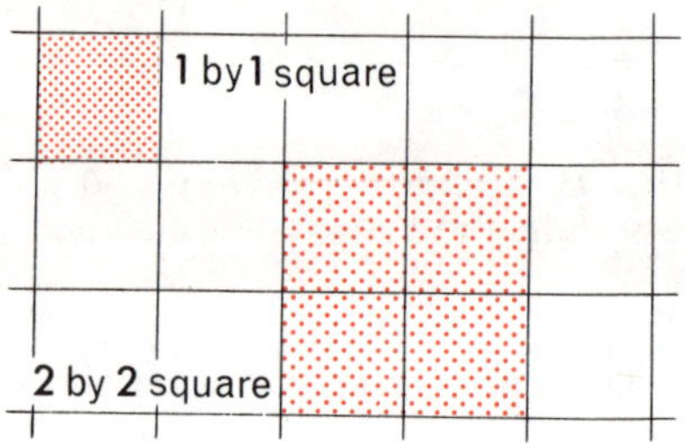

Call these **1 by 1 square** and **2 by 2 square**.

1. Make up some more squares which will help you to complete this table.

Name	1 by 1	2 by 2	3 by 3	4 by 4	5 by 5	6 by 6	7 by 7	8 by 8
Number of Squares	1	4	—	—	—	36	—	—

The numbers in the bottom row of your table are called **square numbers**.

36 is a square number since you can arrange 36 counters into a "square" array.

17 is not a square number since 17 counters cannot be arranged in a square array.

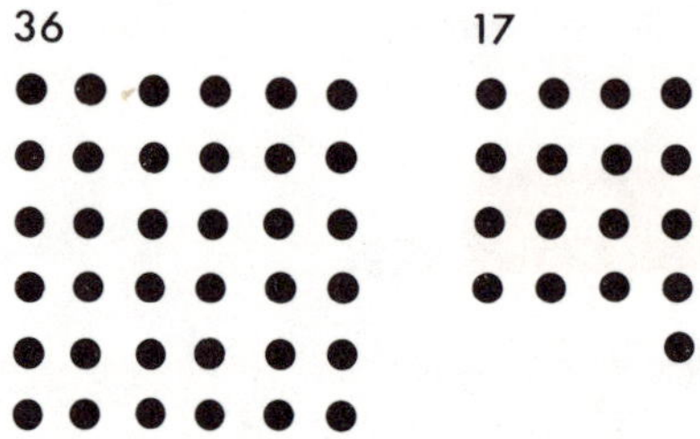

2. This is part of a multiplication square.

a. Copy the framework onto squared paper but make it large enough for 10 by 10.

b. Complete your multiplication square.

c. Shade the squares which contain square numbers.

x	1	2	3	4	5
1	1	2	3		
2	2	4			
3	3				
4				16	
5					

The number 1 is not really a square number. The array for 1 is simply ●

● When a whole number other than 1 is multiplied by itself the product is a **square number**.

This is a 4 by 4 square. Instead of writing 4 by 4 square we can write 4^2

$4^2 = 4 \times 4 = 16$

Using the same notation, we can write $5 \times 5 = 5^2$.

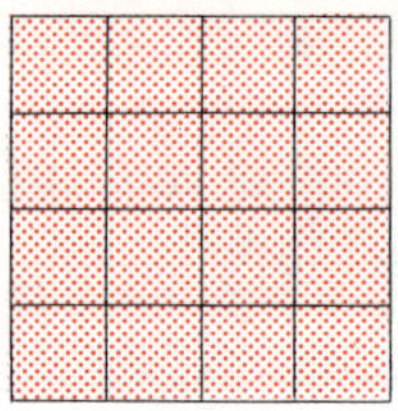

Let's Try

1. Draw diagrams to show squares which are:

   a. 5 by 5 b. 8 by 8 c. 10×10

2. From {38, 49, 81, 120} find out which numbers can be arranged in square arrays.

3. a. Which square number is represented by each of the following?
   6^2, 4^2, 10^2, 11^2

   b. Which is the fifth square number?

   c. Which is the seventh square number?

   d. Complete this table:

Number	1	3	5					11	
Square	1	9		49	64	36	144		400

4. Check that the following statements are correct and then write down the next two lines of the pattern.
   $1^2 = 1$; $2^2 = 1+2+1$; $3^2 = 1+2+3+2+1$

5. Find the replacements for △ and □ in the following:

   a. $5^2 = \triangle \times 5 = 25$ b. $6^2 = \triangle \times \triangle = 36$

   c. $7^2 = \triangle \times \triangle = \square$ d. $\square^2 = \triangle \times \triangle = 64$

6. a. From this series of odd numbers: 1, 3, 5, 7, 9, 11, 13, . . .

$1+3$ $= 4$	$(1+3)+5$ $= 4+5$ $= 9$	$(1+3+5)+7$ $= 9+7$ $= 16$

   Continue working in this pattern.

   b. Your answers are all . . . numbers.

7. a. How many square numbers are there less than 50?

   b. How many square numbers are greater than 50 and less than 100?

Let's Discover

New Ways of Writing Numbers

This is the **numeral** for the number **three hundred and sixty-five**: 365

has three **digits** and can be **expanded** to read 300+60+5.

The numeral for "one thousand and one" is 1001. This is a **four-digit** numeral which is 1000+1 in **expanded form**.

1. This table shows the attendance figures for a series of football matches.

Match	A	B	C	D
Attendance	25000	29000	38000	32000

Copy the axes onto graph paper, using a larger scale. Draw a graph which shows the attendance figures.

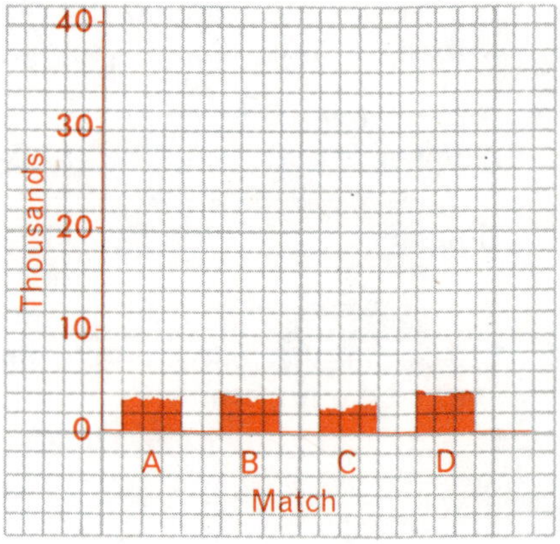

It is unlikely that **exactly** 25000 spectators attended match A.

The number could have been **more**, or **less**, than 25000. The numbers have been given to the **nearest thousand** because we want to know only roughly how many spectators attended.

These numbers would be **rounded off** to 25000 because they are nearer to 25000 than to 24000.

These numbers would be **rounded off** to 25000 because they are nearer to 25000 than to 26000.

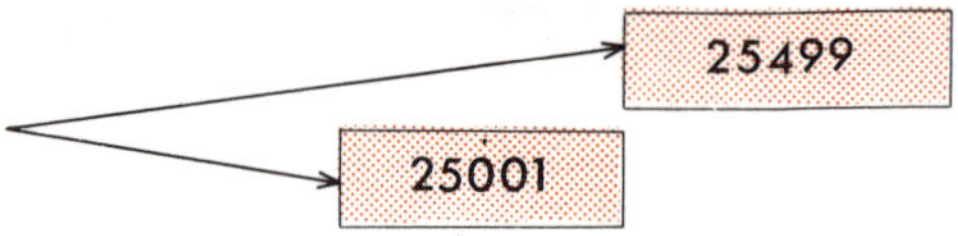

Be generous with 24500. Round it off to 25000 although it is as near to 24000 as it is to 25000.

2. 100000 is a large number. Is 100000 = 10×10×10×10×10?

- Instead of writing 10×10×10×10×10 we can write 10^5.

The "little" 5 names the **power** of 10 and tells us to multiply 10×10×10×10×10.

Using **powers** we can write:

$5 \times 5 = 5^2$; $2 \times 2 \times 2 \times 2 = 2^4$; $4 \times 4 \times 4 = 4^3$

Let's Try

1. Write as four-digit numerals: **a.** one thousand one hundred and eleven; **b.** two thousand and twenty-one; **c.** nine thousand nine hundred and nine.

2. How many thousands are there in:

   a. 27123 **b.** 20123 **c.** 21023 **d.** 2410

3. Write these numbers in expanded form:

   a. 567 **b.** 1432 **c.** 5067 **d.** 2040

4. Write down the numeral which represents:

   a. 300+40+2 **b.** 800+70+4 **c.** 1300+40+6

   d. 10000+100+10 **e.** 4000+300+20+1 **f.** 1000+200+3

5. **a.** What is the greatest number which can be shown by a two-digit numeral?
   b. What is the least number which can be shown by a two-digit numeral?
   c. What is the greatest number which can be shown by a three-digit numeral?
   d. What is the least number which can be shown by a three-digit numeral?

6. Round off the following numbers to the nearest thousand:

   a. 1001 **b.** 999 **c.** 59999 **d.** 60501 **e.** 37409

7. Round off these numbers to the nearest hundred:

   a. 99 **b.** 101 **c.** 753 **d.** 850 **e.** 956 **f.** 11203

8. Write the following as powers:

   a. $2\times2\times2\times2$ **b.** 4×4 **c.** $7\times7\times7\times7$ **d.** 100000

9. Find the replacement for each $\triangle$.

   a. $5^2 = \triangle$ **b.** $5^3 = \triangle$ **c.** $9^2 = \triangle$ **d.** $2^5 = \triangle$

10. Which is the greater 3^4 or 4^3?

11. The sun is $15\times10^{\triangle}$ km from Earth. Find out from your library the replacement for $\triangle$.

Let's Discover **Juno Numbers**

In our number system we count in tens. Column values are:

thousands	hundreds	tens	units
1000	100	10	1
$10\times10\times10$	10×10	10	1

On the planet Juno there is a race of one-armed creatures. These creatures have only one hand and so count in fives. Our system is said to have a **base of ten**. The Juno system has a **base of five**.

A base five system has column values:

$5\times5\times5$	5×5	5	1
125	25	5	1

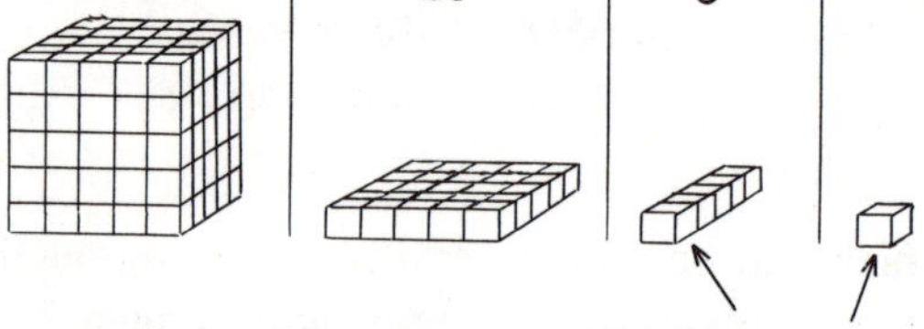

In base five $23_{(5)}$ would be: $2\times5+3\times1 = 13$ in base ten.
Remember to put $_{(5)}$ at the foot of a Juno number to show that you are working in base five.

Juno children count:

one	two	three	four	one–nought	one–one
1	2	3	4	$10_{(5)}$	$11_{(5)}$

1. a. $10_{(5)}$ is one five and no units. How many is this in our number system?
   b. $11_{(5)}$ is one five and one unit. How many is this in our number system?

2. With only two number places we can count up to $44_{(5)}$. How many is this in our system?

   With five groups of five, we need another number column.
   On Juno $100_{(5)}$ is read as **one-nought-nought**.

   $100_{(5)} = (1\times25)+(0\times5)+(0\times1) = 25_{(10)}$

3. Write these Juno numbers in words and then change them to base ten.
   $102_{(5)}$; $104_{(5)}$; $111_{(5)}$; $121_{(5)}$; $200_{(5)}$; $201_{(5)}$

- A base ten number can be changed to base five by splitting it up into: . . ., 25s, 5s, 1s.

For example, $73_{(10)} = (2\times25)+(4\times5)+3$

and so $73_{(10)} = 243_{(5)}$

Let's Try

1. Write these Juno numbers in words:

   a. $3_{(5)}$ b. $10_{(5)}$ c. $12_{(5)}$ d. $21_{(5)}$ e. $100_{(5)}$ f. $202_{(5)}$

2. a. b. c. d. e.

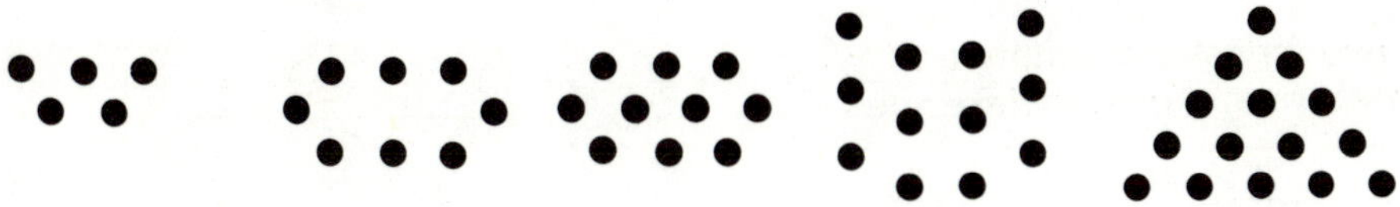

   Count the number of dots in each pattern in base ten and then write the Juno numbers.

3. Copy this table and then use it to help you change Juno numbers to base ten.

base five				base ten			
twenty-fives	fives	units		H	T	U	
	4	3	$(4\times5)+(3\times1)$		2	3	$43_{(5)} = 23_{(10)}$

   a. $12_{(5)}$ b. $22_{(5)}$ c. $44_{(5)}$ d. $120_{(5)}$ e. $211_{(5)}$ f. $403_{(5)}$
   g. $444_{(5)}$

4. The column values for Juno numbers are . . . , 25, 5, 1.
   What is the missing number?

5. Tom arrived on Juno with twenty-three marbles and decided that each of his friends could have five marbles.

   a. How many friends could each receive five marbles? How many marbles had Tom left?
   b. Is it true that $23_{(10)} = 43_{(5)}$?

6. Change these base ten numbers to base five:

   a. 13 b. 16 c. 24 d. 30 e. 31 f. 69 g. 76
   h. 126 i. 256

7. Rewrite these sentences (base ten) using base five numbers:

   a. That lorry has 8 wheels.
   b. It took me 90 minutes.
   c. The price is 77p.
   d. My grandfather is 82.
   e. John has saved 100p.
   f. Nessie has 80 legs.
   g. The race was over 130 km.
   h. That table is 250 cm long.

Let's Discover

Some Line Graphs

Find out the name of this type of bicycle from your library.

Unlike modern bicycles this one has wheels which are different sizes.

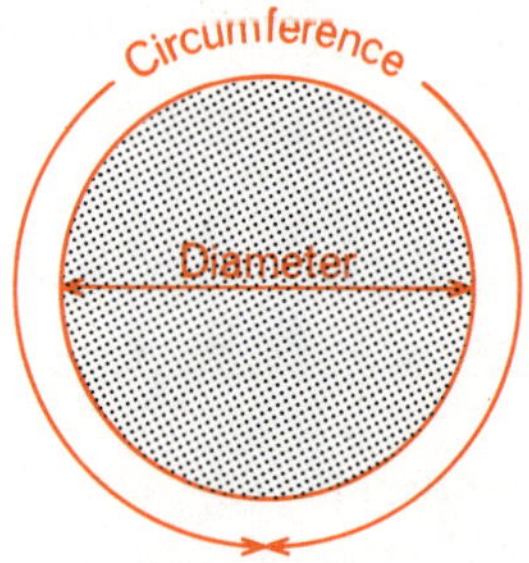

The two wheels have different **diameters** and **circumferences**.

1. a. Which wheel has the larger diameter?
   b. Which wheel has the larger circumference?

Here is a simple way of measuring the circumference of a wheel.
Wrap a strip of paper tightly round the wheel and push a pin through the overlap.
Remove the pin and paper and measure the distance between the pin holes.

2. a. Collect 6 wheels or circular lids of different sizes.
   Number them in order of size with the smallest as number 1.
   b. Use a pair of calipers to measure the diameter of each wheel.
   c. Measure the circumference and complete a table like this:

circle	1	2	3	4	5	6	
diameter (D)			7				cm
circumference (C)			22				cm

   d. Draw a pair of axes and graph the pairs (D, C).

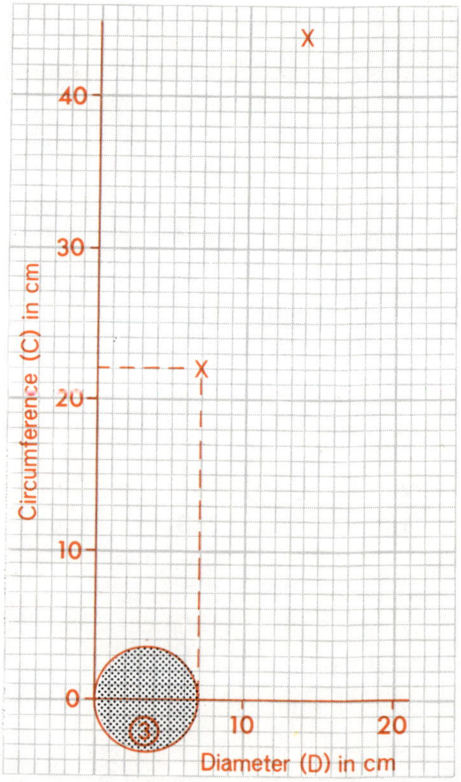

3. Refer to your graph of pairs (D, C).
   The points representing the pairs should lie on a straight line.
   Check that yours do and then draw a straight line through them.

Here is a graph showing how Old Father Tom travelled when he went for a bicycle ride.

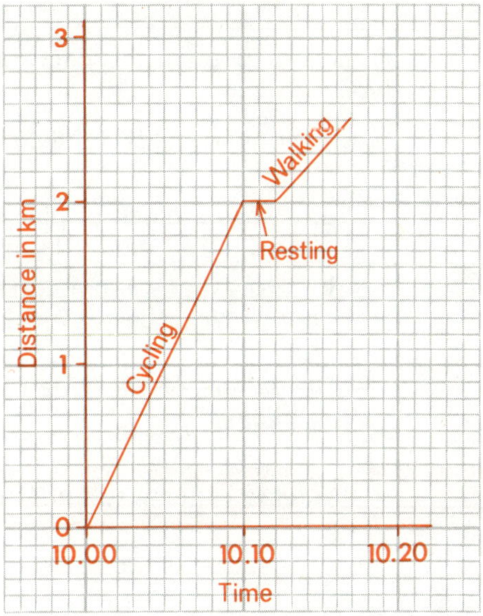

4. Check on the graph that he:

a. cycled 2 km in 10 min

b. rested for 2 min

c. walked $\frac{1}{2}$ km in 5 min

Between 10.00 and 10.10 it is assumed that Tom cycled at a steady speed.
He travelled 2 km in 10 min so in 60 min he would have travelled 6×2 km.
Therefore his speed is 12 km per hour.

d. Check that his walking speed is 6 km per hour.

- Remember that the steepness of a line graph shows how the values plotted are changing.

Let's Try

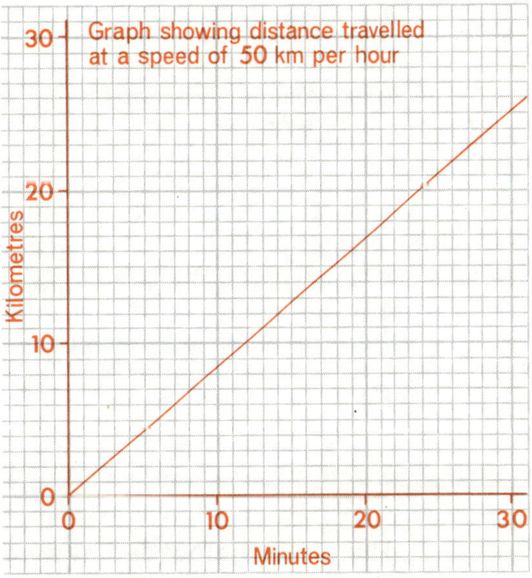

1. Here is a graph showing distance travelled in various times at a speed of 50 km per hour.

a. At a speed of 50 km per hour, how far will I travel in:
30 min; 12 min; 24 min; 18 min?

b. At a speed of 50 km per hour, how long will it take to travel:
25 km; 5 km; 20 km?

c. Draw a similar graph showing the distance travelled in various times at a speed of 60 km per hour.

d. On the same axes draw a graph showing the distance travelled at a speed of 6 km per hour.

2. Here is a table showing the approximate population of England, Scotland and Wales.

Year	1911	1921	1931	1951	1961	1971	
Population	41	43	45	49	51	53	millions

Draw a graph showing this information.

Let's Discover **Parts of a Litre**

This is a spoon with which nurse gives you medicine.

Each spoonful holds 5 millilitres.

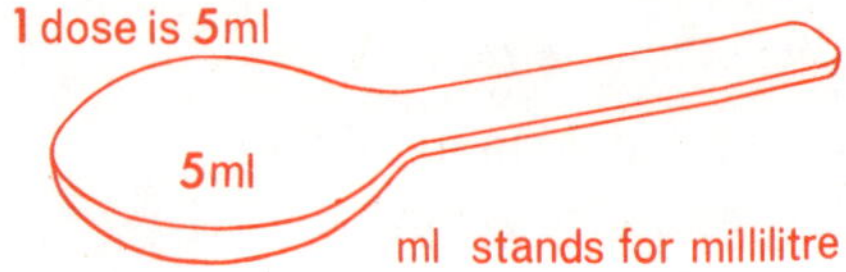

1. **a.** Make a block of plasticine 5 cm by 1 cm by 1 cm.

b. How many 1-cm cubes can this block be cut into?

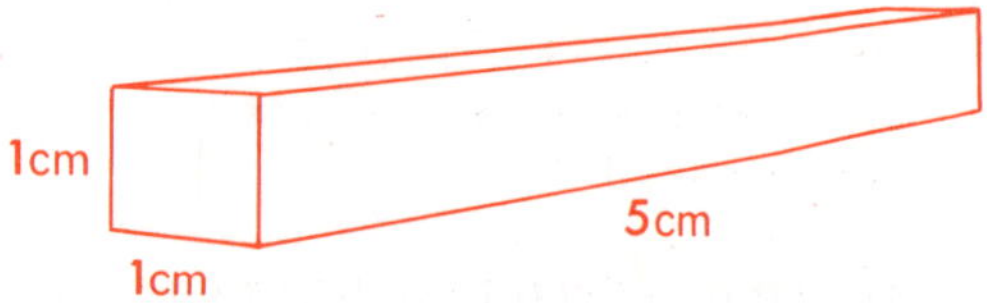

The volume of medicine in the spoon is the same as the volume of the block of plasticine.

- 1 cubic centimetre (cm^3) is the same volume as 1 ml.

2. Complete the following sentence.
The volume of this box
$= 10 \times 10 \times 10$ cm^3
$= \ldots$ cm^3

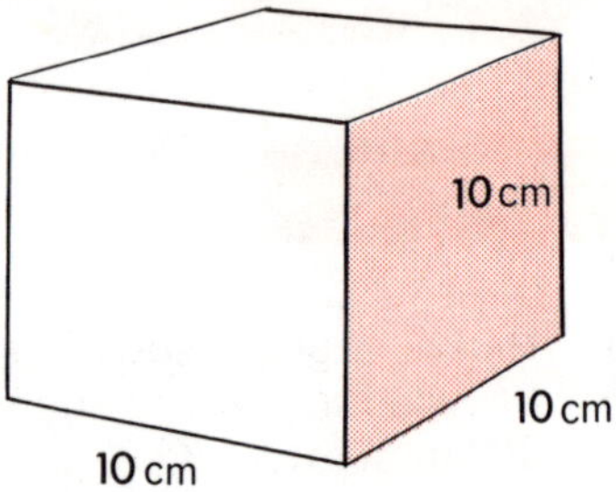

The volume of this box is **1 litre** or 1000 cm^3 or 1000 ml

3. From this bottle nurse can give **4 doses** per day for **5 days**, which is 20 doses altogether. Each dose is 5 ml.
Fill in the missing numbers.

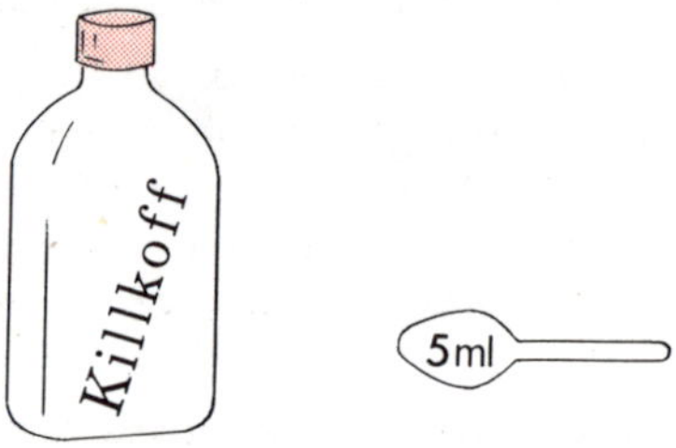

The bottle contains . . . doses,
which is $20 \times$ . . . ml
= 100 ml

The medicine bottle has a capacity of 100 ml.

4. Complete this working.
The total **capacity** of 10 of these bottles
$= 10 \times 100$ ml = . . . ml = . . . litre
so 100 ml is $\frac{1}{10}$ of a litre

The nurse has a bottle of disinfectant which holds $\frac{2}{5}$ litre.

$\frac{2}{5}$ litre $= \frac{4}{10}$ litre $= 400$ ml

Let's Try

1. Here is a 1-litre can of cooking oil. Fill in the missing numbers.

   a. A 1-litre can contains . . . ml.

   b. A $\frac{1}{2}$-litre can contains . . . ml.

   c. A $\frac{1}{4}$-litre can contains . . . ml.

   d. A $\frac{3}{4}$-litre can contains . . . ml.

2. Dad's oil can holds 200 cm^3.

   a. How many times can he fill it from a 1-litre tin of oil?

   b. If the oil costs 50p per litre, how much does it cost to fill the can?

3. These milk cartons contain 250 ml.

   a. How many ml are contained in 4 cartons?

   b. What fraction of a litre is contained in each carton?

   c. A milk float carries 400 cartons. How many litres of milk is this?

   d. The cost of 1 carton is $3\frac{1}{2}$p. What is the price of milk per litre?

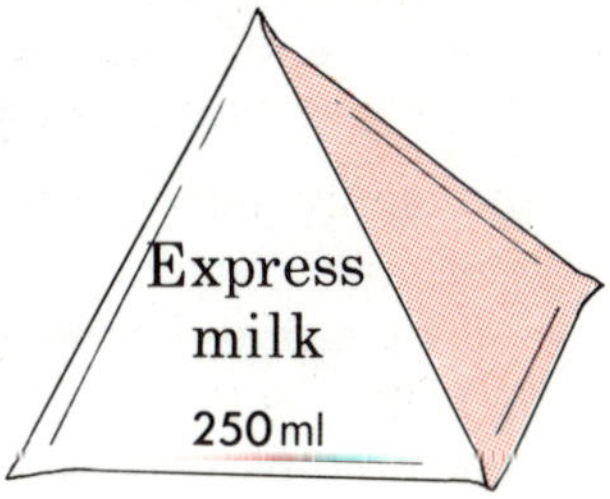

4. I travel 1500 km a month in my car. A litre of petrol takes me 10 km.

   a. How much petrol do I use each month?

   b. What is my monthly petrol bill if petrol costs 8p per litre?

5. Adjust a water tap so that it drips 1 drop about every 5 seconds.

   a. Estimate how many drops will fill $\frac{1}{10}$ litre.

   b. Check how many drops make $\frac{1}{10}$ litre. From your answer work out how many drops make $\frac{1}{2}$ litre and 1 litre.

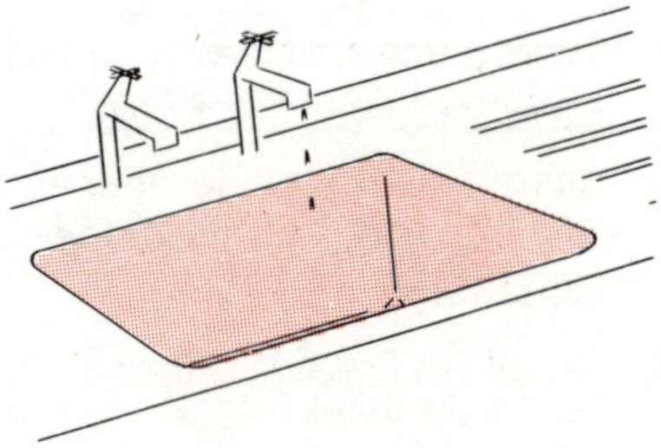

Let's Discover

A New Kind of Graph

Three boys, Tom, Ben and Jack sat down to eat a pie. When it had all gone they were wondering who had eaten the most. Tom said, "I have drawn a picture of the pie and coloured it to show how much each of us has eaten."

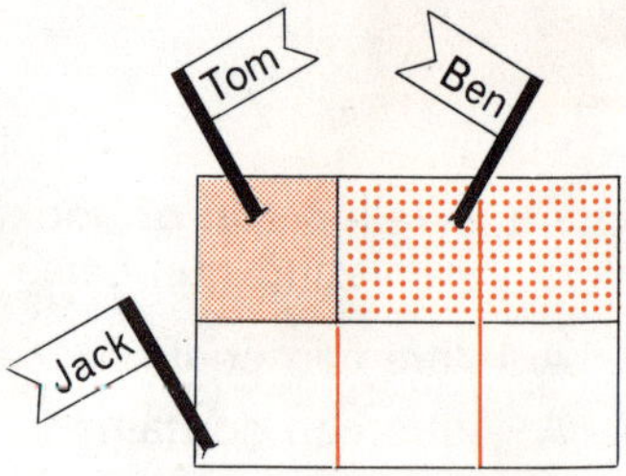

Chart showing how the pie was eaten

1. a. Who ate the most pie? What fraction did he eat?

   b. What fraction of the pie was eaten by Tom?

The picture of the pie shows us clearly what fraction of the pie was eaten by each boy. A picture like this is called a **pie chart**.

2. Here is a pie chart representing a box of chocolates and showing us what part of the chocolates was eaten by each of four girls.

   a. Who ate the most? Who ate the least?

   b. Who ate $\frac{1}{8}$ of the chocolates? Betty ate . . . of the chocolates.

   c. There were 80 chocolates in the box so Mary ate $\frac{1}{4}$ of 80 = . . . chocolates.

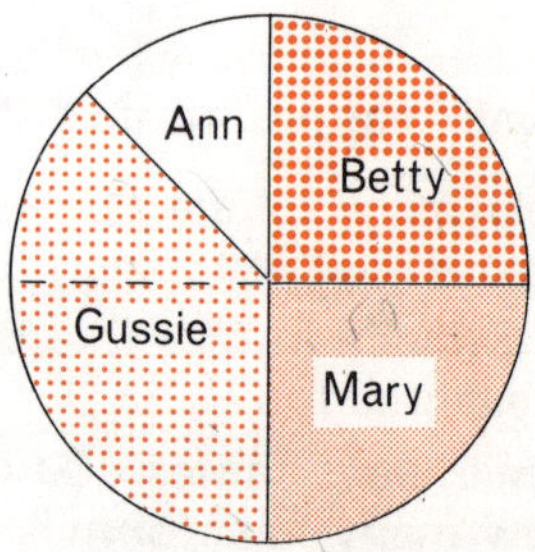

Chart showing how the chocolates were eaten

   d. Find out how many chocolates were eaten by each girl? Add your four numbers together to check that the total is 80.

Jenny Copper saved her money to buy presents. She spent:

£2 on Mother, £2 on Father,
£1 on each of her 3 brothers
and put the £1 she had left in the bank.

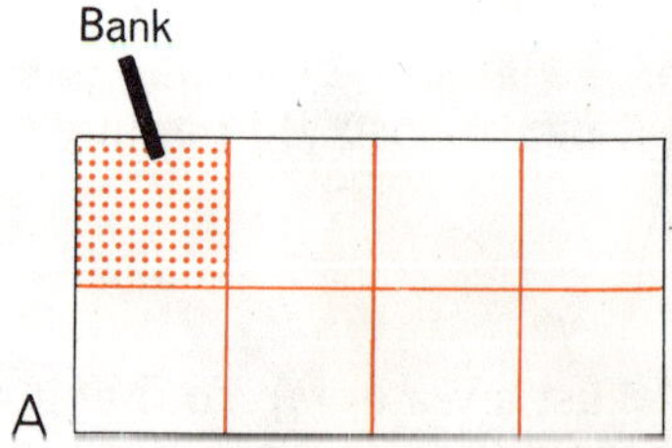

Chart showing how Jenny used her **£8**

3. a. How much had Jenny saved altogether?
   £2+£2+(3×£1)+£1 = £ . . .
   Jenny spent $\frac{2}{8}$ of her money on Mother and $\frac{2}{8}$ of her money on Father.
   What fraction did she:

   b. spend on her 3 brothers?

   c. put in the bank?

4. a. Diagram A shows a rectangular pie chart on which we can show how Jenny used her £8. Copy and complete the chart.

b. Cut out a circle of paper and fold it into eighths. Use your circle to copy and complete diagram B showing how Jenny used her money.

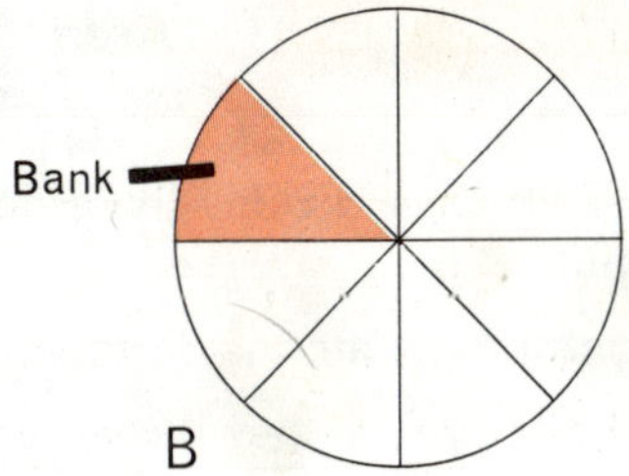

Let's Try

1. A class of 40 pupils was asked to vote for their favourite sport. Copy this pie chart which shows how the children voted.
What fraction of the class voted for:

a. cycling   b. cricket   c. swimming

How many children voted for:

d. swimming   e. cricket   f. cycling

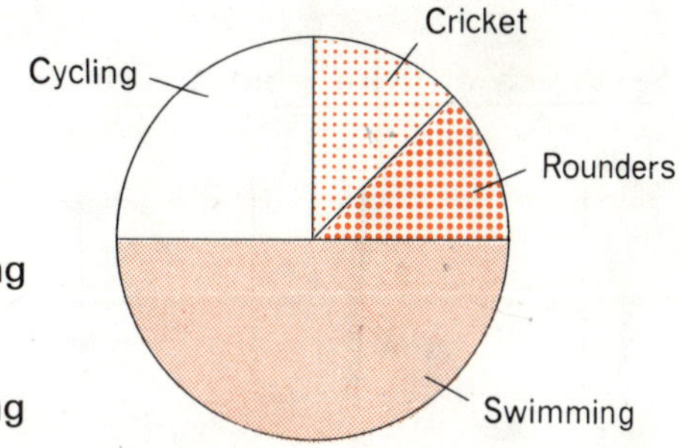

2. Draw a square 8 units by 8 units and use it to show the information from question **1** on a pie chart.

3. This pie chart shows the results for team A which played 32 matches.
What fraction and how many games were:

a. won (W)   b. drawn (D)
c. lost (L)
d. Team B won 4, drew 12 and lost 8 matches.
Draw a circular pie chart and a rectangular pie chart to show the results for team B.

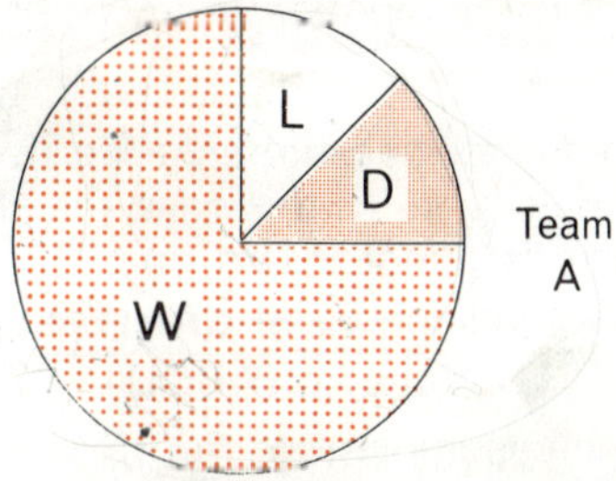

4. Here is a plan of the ground floor of a house.
What fraction of the floor is used for the:

a. kitchen (K)   b. playroom (P)
c. hall (H)   d. dining-room (D)
e. lounge (L)
f. Work out the total area of the floor and the area of each room in square metres (m^2).
g. On a circular pie chart show how the ground floor is used.

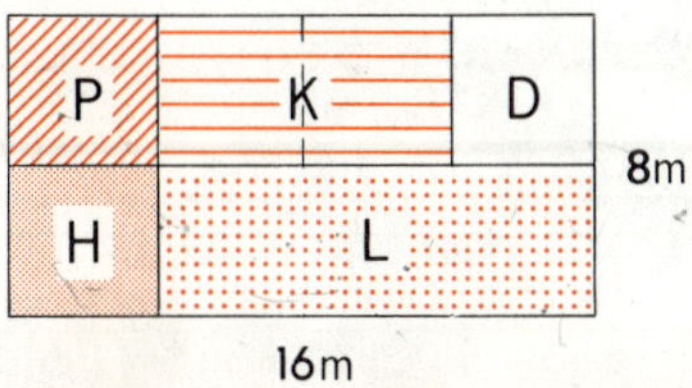

Let's Discover **Area**

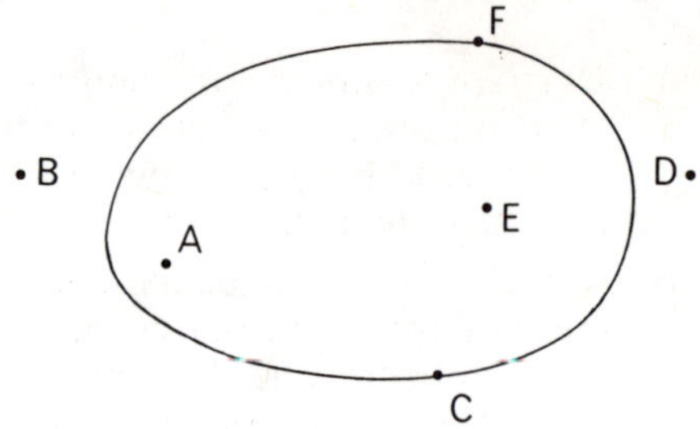

This closed curve divides the plane into two regions.

1. Which points are inside the curve?
2. Which points are on the curve?
3. Which points are outside the curve?

   Here are two enclosed regions. One is triangular and the other is circular.

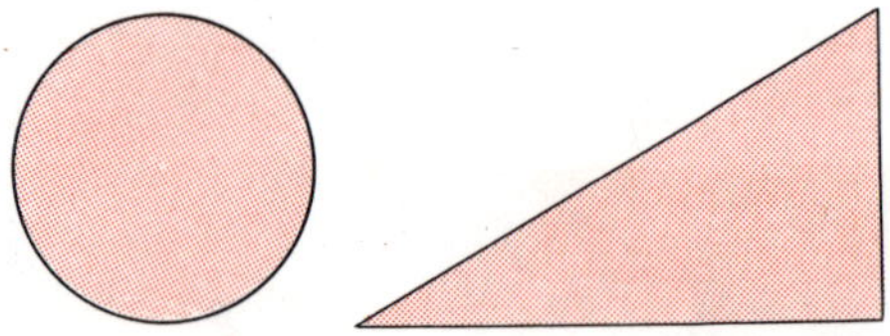

4. Guess which of these enclosed regions is the larger.
5. Which of these regions is the largest and which is the smallest?

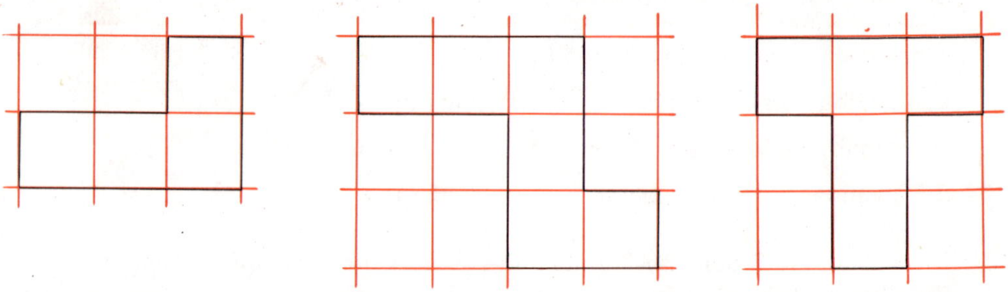

- Area is the measurement of a region.

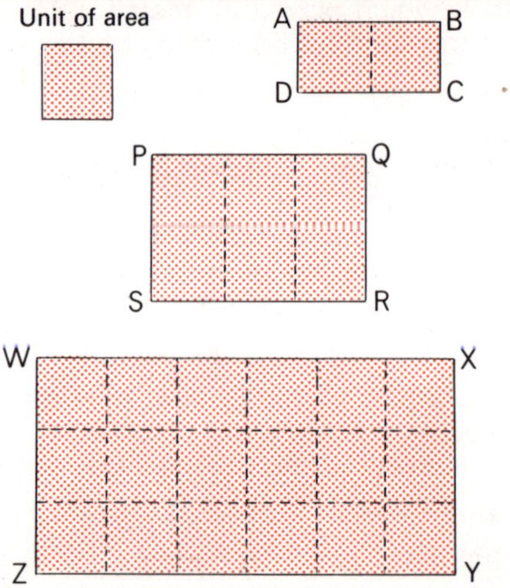

In order to measure area we need a standard unit. In this case the shaded square is our standard unit.

The measurement of ABCD is 2 units.
We write: m(ABCD) = 2 units

The measurement of PQRS is 6 units.
We write: m(PQRS) = 6 units

The area of WXYZ is 18 units.
There are 3 units along one side and 6 units on the other.

$m(WXYZ) = \triangle \times \triangledown$
$= 18$ units
so $\triangle = \ldots$ and $\triangledown = \ldots$

- A standard unit for measuring area is a centimetre square.

Cut out a centimetre square and check that you need six such squares to cover this rectangle.

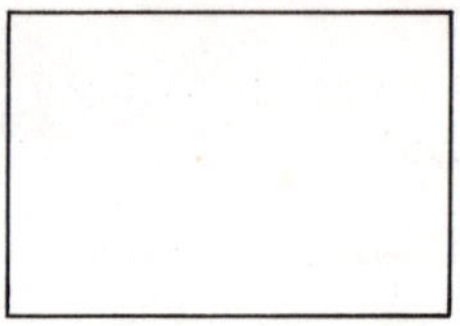

Let's Try

1. Complete the following equations using  as the unit of area.

   a. 2×6 = . . .
   m(PQRS) = . . . units

   b. 4×3 = . . .
   m(LMNO) = . . . units

   c. . . . × . . . = 10
   m(ABCD) = 10 units

   d. . . . × . . . = . . .
   m(EFGH) = 24 units

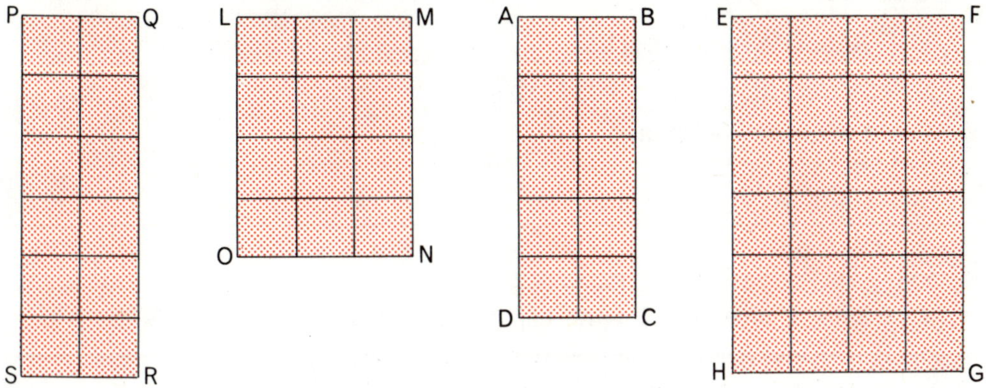

2. This picture shows a page of a book of gift stamps which are being given away by a shopkeeper.

   a. How many rows are there on the page?

   b. How many columns are there on the page?

   c. How many stamps are there on the page?

   d. What is the area of the unused region if each stamp is a unit of area?

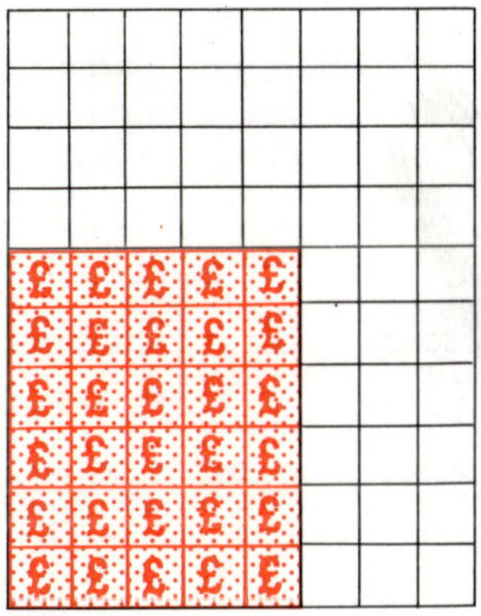

3. Here is a picture which I am going to cover with glass.

   a. How many cm long is the picture?

   b. How many cm wide is the picture?

   c. What is the area of the picture in cm^2?

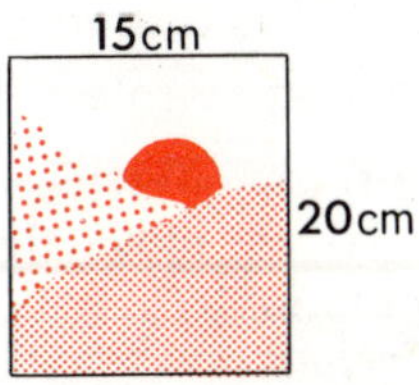

4. a. This rectangular region has been accurately drawn. Find its area.

   b. What is the area of the red region?

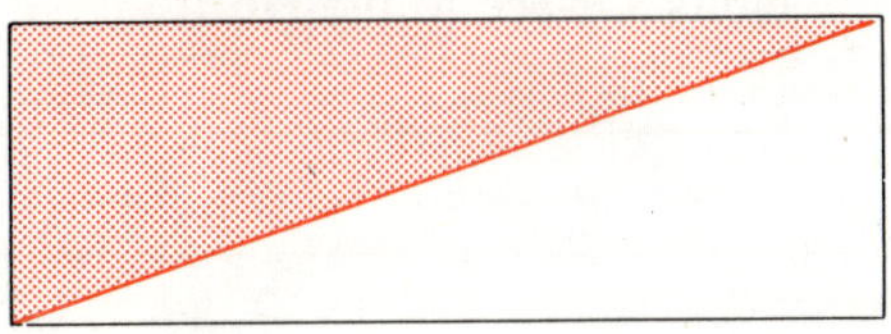

Let's Discover

More Base Ten Numbers

We are now going to have another look at decimal fractions.

1. Here is a chart which shows place and face values.

   Check that the total of the face values is 234·56.

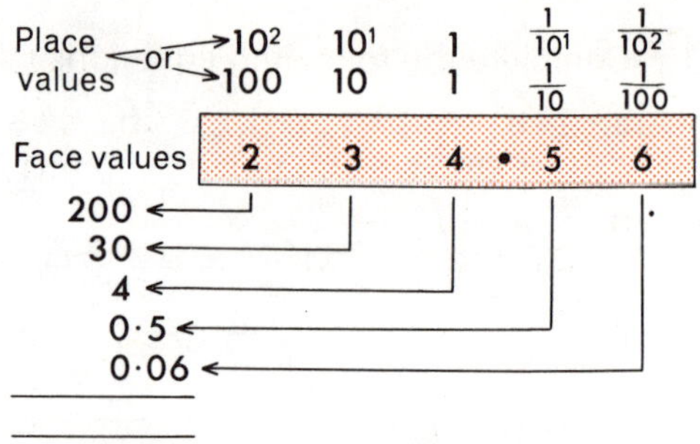

2. Could we write 234·56 as $234+\frac{5}{10}+\frac{6}{100}$?

3. Complete the following:

   a. 3 tens, 2 ones and 1 tenth can be written as 32· . . . or $32\frac{1}{10}$

   b. 3 tens and 2 tenths can be written as 30·2 or . . .

   c. 3 tens and 2 hundreths can be written as . . . or . . .

4. Earlier in this book you learned how to write numbers in expanded form. What are the missing column values in the following?
   214 = (2× . . .)+(1× . . .)+(4× . . .)

5. Write these decimals in expanded form:

   a. 12·34 = (1× . . .)+(2× . . .)+$(3\times\frac{1}{10})+(4\times\frac{1}{100})$

   b. 123·4 = . . . **c.** 1·23 = . . . **d.** 1·32 = . . .

6. Mary was asked to write 125p in £.

   a. She wrote £12·5. Is this answer 10 times too large?

   b. She tried again and wrote £1·25. Is this answer correct?

£1·25×10 = £12·5
To **multiply** by 10 we move the digits 1 place to the **left**.

and

£1·25×100 = £125
To **multiply** by 100 we move the digits 2 places to the **left**.

£12·5÷10 = £1·25
To **divide** by 10 we move the digits 1 place to the **right**.

and

£125÷100 = £1·25
To **divide** by 100 we move the digits 2 places to the **right**.

Let's Try

1. Write the **place value** and the **face value** of the 5 in:

   a. 25·1 b. 21·5 c. 4·05 d. 15·32 e. 13·52 f. 1·352

2. Write the correct symbol: <, >, or = in each □

   a. 2·3 □ 3·2 b. 0·6 □ $\frac{6}{10}$ c. 0·05 □ 0·5 d. 0·23 □ 0·32

3. Add the two decimals.

   a. $\begin{array}{r} 12·37 \\ +27·44 \\ \hline \\ \hline \end{array}$ b. 29·9+14·32 c. 147·05+9·7 d. 256·7+63·07 e. 99·99+999·9

4. Complete the following:

   a. $\begin{array}{r} 67·36 \\ -16·43 \\ \hline \\ \hline \end{array}$ b. 82·82−22·95 c. 68·3−9·23 d. 427·45−126·7 e. 888·8−99·99

5. Use the method of moving digits to find:

   a. 1·25×10 b. 1·25×100 c. 1·25×1000 d. 1·25÷10
   e. 47·02÷10 f. 47·02÷100 g. 47·02×10 h. 47·02×100

6. Write the next two numbers in these series:

   a. 0·3, 0·6, 0·9, . . . , . . . b. 0·25, 0·3, 0·35, . . . , . . .
   c. 0·11, 0·15, 0·19, . . . , . . .

7. The decimal point is missing from each of the following numbers. Put the points in so that each 5 has a value of $\frac{5}{10}$.

   a. 1453 b. 105 c. 12051 d. 6050

8. Ann is saving 2p per day from her pocket money. How much will she save in:

   a. 100 days b. 200 days

9. Tom is managing to save $2\frac{1}{2}$p per day from his pocket money. How much will he save in:

   a. 10 days b. 100 days

10. The petrol tank of my car holds 20·5 litres. If the tank is only $\frac{1}{10}$ full how much petrol is there in the tank?

Let's Discover

Multiplication and Division of Money

Here is a box of eggs.

1. Complete the following:

   a. Each egg costs . . . p

   b. 10 boxes of eggs cost . . . p
   or £ . . .

On page 28 you learned how to multiply by 10 and how to divide by 10.

When you have to find the answer to a problem you need to think about which operation: +, ×, −, or ÷, you are going to use.

2. Are you going to **multiply** or **divide** to find the answer to:

a.

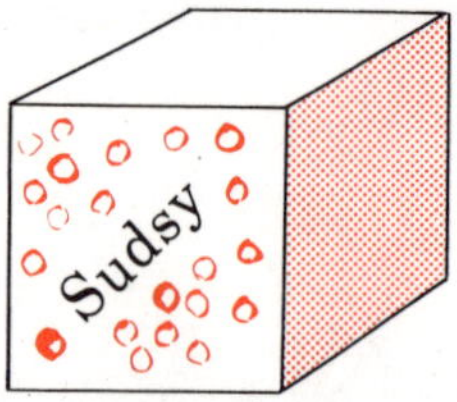

9 boxes of Sudsy cost £0·90
What is the cost of 1 box?

b.

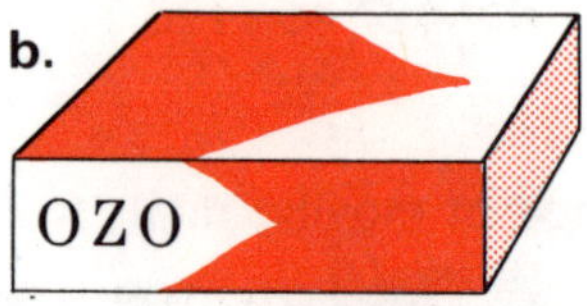

1 box of Ozo costs 12p
What is the cost of 4 boxes?

To find the cost of 8 Tums if 1 costs £0·23, we multiply.

$$\begin{array}{r} £ \\ 0{\cdot}23 \\ \times \quad 8 \\ \hline 1{\cdot}84 \\ \hline \end{array}$$

To find the cost of 1 Banger if the cost of 7 is £1·26, we divide.

$$\begin{array}{r} £ \\ 0{\cdot}18 \\ 7\overline{)1{\cdot}26} \end{array}$$

Think whether your answer is about right. Write statements such as:

8 Tums cost £1·84   1 Banger costs £0·18

to show your answer.

Sometimes money problems will involve halfpennies.

3. Complete the following:

   a. 8 halfpennies make . . . p   b. 13 halfpennies make . . . $\frac{1}{2}$p

   When multiplying amounts involving halfpennies, work like this:

$$\begin{array}{r} £ \\ 0{\cdot}06 \\ 7 \\ \hline 0{\cdot}42 \\ \hline \end{array}$$

and $\frac{1}{2}$p × 7 = 3$\frac{1}{2}$p

so £0·06$\frac{1}{2}$ × 7 = £0·42 + 3$\frac{1}{2}$p
= £0·45$\frac{1}{2}$

Let's Try

1. Practise your × by completing these examples:

a.	b.	c.	d.	e.
£	£	£	£	£
0·40	1·25	2·07	0·04½	1·02½
× 6	× 7	× 8	× 9	× 20

2. Practise your ÷ by completing these examples:

a.	b.	c.	d.
£	£	£	£
5)5·25	6)1·32	12)1·32	8)9·20

3. A shopkeeper is ordering new stock for his shop.
Here is a page from his catalogue.

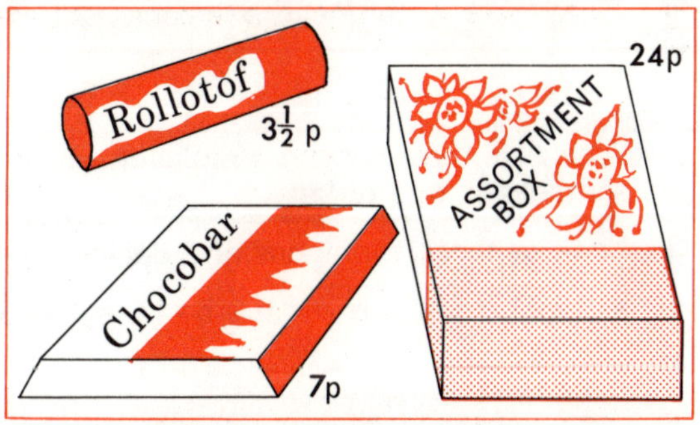

Write the cost of:

a. 100 Chocobar
b. 10 Assortment Box
c. 10 Rollotof
d. 100 Rollotof

Copy and complete these bills:

e. 20 Rollotof @ 3½p each
15 Chocobar @ 7p each
2 Assortment Box @ 24p each — £

f. 200 Rollotof @ 3½p each
150 Chocobar @ 7p each
20 Assortment Box @ 24p each — £

4. Here are some problems which can be worked out in your head.
Try to use a short method to find the cost of:

a. 4 articles at £0·99 each
b. 5 articles at £1·98 each
c. 6 bags of fuel at 54p each

5. Complete these bills for the post office assistant:

a. 16 stamps at 2½p each
1 parcel at 35p
20 stamps at 10p each — £

b. Telephone bill:
200 units at 1p each
4 trunk calls at 25p each
Rental £6 — £

c. 4 reg. envelopes @ 12p each
1 TV licence @ £6
14 stamps @ 2½p each — £

Let's Discover **Solution Sets**

You have already learned how to use the symbols: $<$ $>$ $=$ $\neq$

1. a. What do the symbols $<$, $>$, $=$ and $\neq$ mean?
   b. Which is the greater number, a or b, if:
   $a < b$ and $b > a$?

2. Some number sentences need a replacement. What is the whole number replacement for $\triangle$ in these sentences?

   a. $\triangle + 3 = 9$ b. $14 - \triangle = 9$ c. $\triangle \div 6 = 7$ d. $\triangle \times 3 < 14$
   Did you notice that part **d** had more than one replacement?

- Sentences which need a replacement are called **open sentences**.

When we have chosen a replacement for $\triangle$, we can then decide whether the sentence is true or false.

3. Look at these sentences and decide whether they are true or false:

   a. $2 + 17 = 17 + 2$ b. $5 \times 14 = 14 \times 5$ c. $24 \div 12 = 12 \div 24$

- Remember that these sentences are called **closed sentences** because they do not need a replacement.

4. Here is an open sentence which needs a replacement: $\boxed{\triangle \times 3 < 14}$
   Using {2, 4, 6, 8, 10} as our **universe**,

   a. Does $\triangle = 2$ make the sentence true? b. Does $\triangle = 4$ make the sentence true?

   c. Check that no other element from the universe makes the sentence true.
   {2, 4} is called the **solution set** to the sentence for the given universe.
   Each element of {2, 4} is represented by a X on this number line.

5. Here is a graph on a number line.

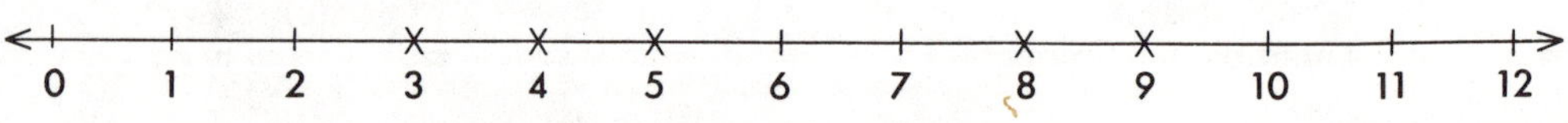

Is each element of {3, 4, 5, 8, 9} shown on the graph?

- A set of numbers can be graphed on the number line.

Let's Try

1. Find the whole number for $\triangle$ which makes each sentence true.

   a. $8 \times 3 = \triangle \times 8$ b. $9 + \triangle < 11$ c. $\triangle - 9 = 7$
   d. $9 - \triangle > 7$ e. $7 \times \triangle = 49$ f. $\triangle \div 8 = 9$

2. Write $<$, $>$ or $=$ to replace $\square$ to make the sentences true.

   a. $4 \square 3$ b. $3 \square 4$ c. $7 \times 6 \square 6 \times 7$
   d. $9-5 \square 10-5$ e. $5+5+5 \square 555$ f. $(12 \div 2) \square 2 \times 3$

3. Select the correct sign for $\square$ from $\{+, -, \times, \div\}$ as a replacement to make the sentences true.

   a. $20 \square 12 = 8$ b. $13 \square 4 = 4 \times 13$ c. $10 \square 4 = 2\frac{1}{2}$
   d. $6 \times 6 = 9 \square 4$ e. $6 \square 6 < 12$ f. $2 \square 2 \neq 1$

4. Here are some sentences. Which ones are closed sentences, and which closed sentences are false?

   a. $50 - 25 = 25$ b. $40 - \triangle > 10$ c. $8 \times 6 = 4 \times 12$
   d. $15 \div 5 = 3$ e. $10 \times 5 < 0$ f. $37 + 14 = 53$

5. Rewrite these word sentences in number sentences. Find the number I thought of in each sentence.

   a. I thought of a number, $\triangle$, added five to it and got fifteen.
   b. I thought of a number, $\triangle$, subtracted ten from it and got twenty.
   c. I thought of a number, $\triangle$, divided it by two and got ten.

6.

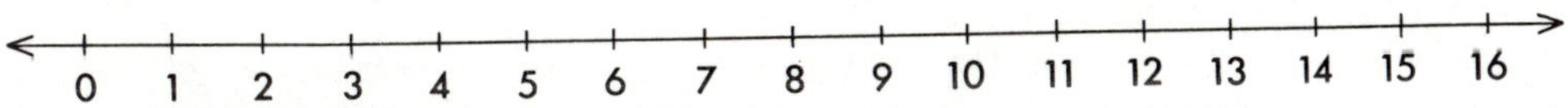

   a. Copy this number line for each of the following sets.
   $A = \{3, 6, 9\}$; $B = \{0, 1, 2\}$; $C = \{0, 1, 3\}$; $D = \{9, 12, 14\}$
   b. Graph each set on a number line.
   c. Which of the sets A, B, C, D is the solution set to $2 \times \triangle < 5$ if $\triangle$ is replaced by a number from your number line?

Let's Discover — Some Lengths

John is trying to measure the length of his shadow. He made a mark on the floor at the tip of his toes.
He is now trying to make a mark at the far end of his shadow.

1. a. Measure your own shadow length. Don't do it like John, get a friend to help.
   b. Measure your own height.
   c. Draw a column graph showing height for 10 children in your own class ▨ and shadow lengths ▧ (Shadows must be measured at about the same time.)
   d. Who has the longest shadow and who is the tallest?
   e. Who has the shortest shadow and who is the smallest in height?
   f. Make up some more questions about your graph.

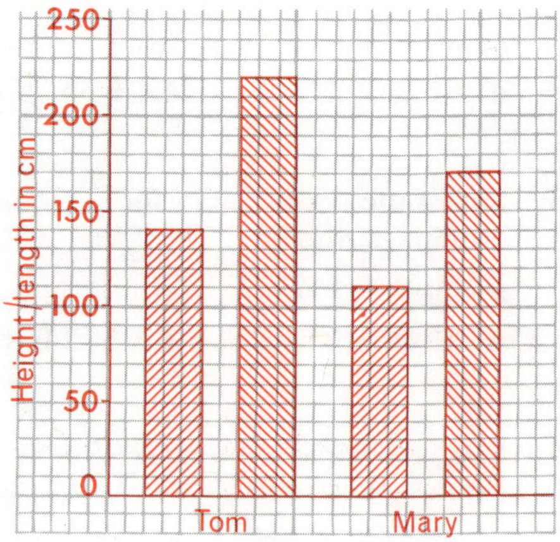

Did you measure your height and shadow length in metres and cm, or just cm?

My height is 1 m 65 cm
1 m 65 cm is the same height as 100+65 cm
1 m 65 cm = 165 cm

On page 28 you learned how to divide by 100.
This is how we change cm to m.

165 cm = $\frac{165}{100}$ m = 1·65 m

Here is another example:

65 cm = $\frac{65}{100}$ m = 0·65 m

John has stopped chasing his shadow.
He is wondering how tall the baby giraffe is.

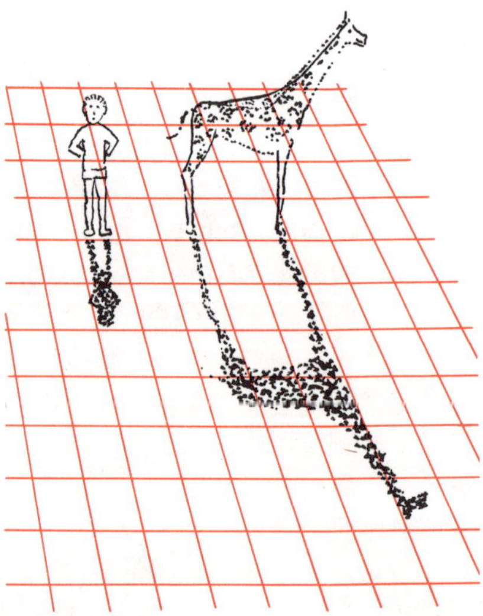

2. Look at the picture and check that the giraffe's shadow is 3 times as long as John's shadow.

- Since the giraffe's shadow is 3 times as long as John's shadow, the giraffe is 3 times as tall as John.

3. Complete this working.
   John is 1 m 5 cm or 1·05 m tall,
   so the giraffe is 1·05 m × 3 = . . . m tall.

Are you good at guessing heights? If you know your own height it will help you guess the height of a lamp-post.
Stick a piece of coloured paper on the post at your height.
Stand back from the post and try to guess how many times your height makes the same height as the lamp-post.

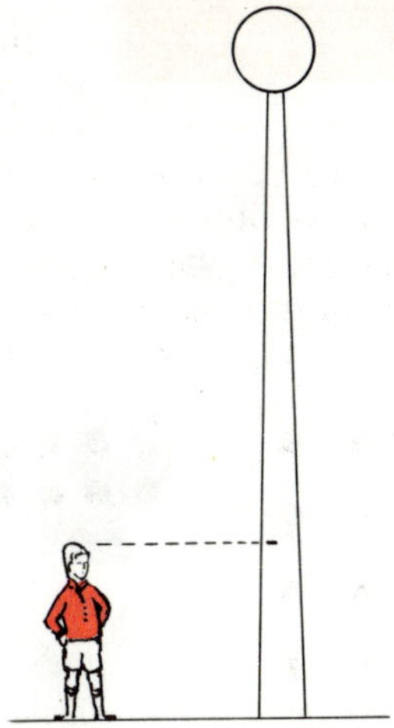

4. The lamp-post is 4 × the height of Tom.

a. If Tom is 1 m tall the lamp-post is . . . m tall.

b. If Tom's shadow is 1·5 m long the shadow of the lamp-post will be 4 × 1·5 m = . . . m long.

Let's Try

1. Choose a sunny day for question **1**.

a. Set a 1-metre stick upright in the ground. Measure the shadow every half-hour and complete a graph showing:

The length of the shadow of a 1-metre stick during the day.

b. When is the shadow shortest? Where is the sun at this time?

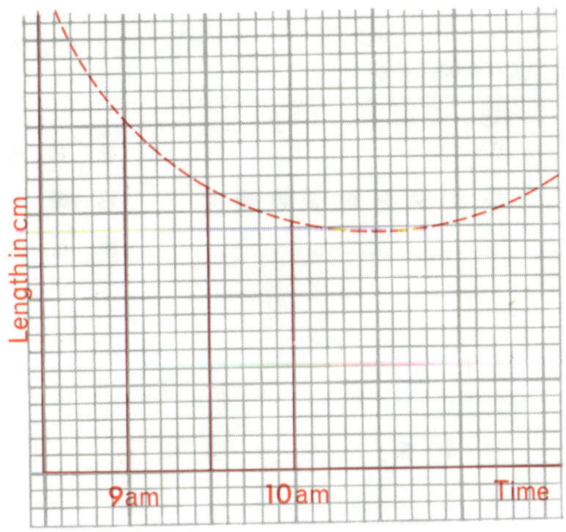

2. Change each of the following to a decimal of a metre.

a. 1 m 20 cm **b.** 1 m 2 cm **c.** 2 m 25 cm **d.** 2 m 52 cm

3. Change the following to m and cm.

a. 1·25 m **b.** 2·56 m **c.** 2·65 m **d.** 3·05 m

4. Estimate the height of:

a. your school building **b.** a tree **c.** a telegraph pole

5. Complete the following:

a.

m	cm
4	70
+3	65
___	___

b.

m
2·75
−1·96
___

c.

m	cm
2	30
×	5
___	___

Let's Discover

Rectangular Numbers

On page 14 you were asked to shade square numbers on a multiplication square.
Some square numbers were shaded more than once.
For example, 4 was shaded at: 4×1 1×4 2×2
The square number 16 was shaded at:
4×4 8×2 and 2×8 16×1 and 1×16

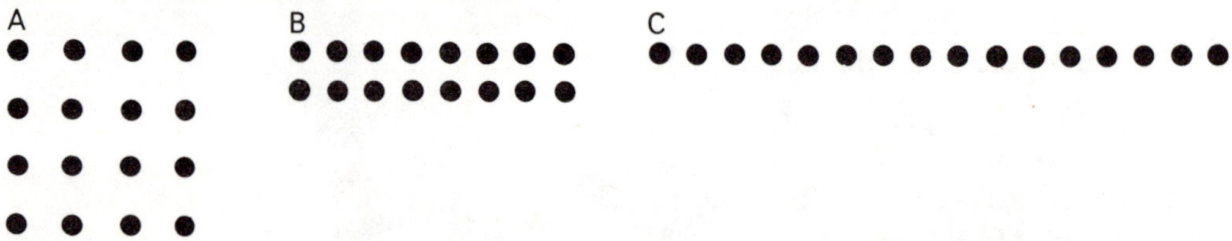

Pattern A is a **square**, pattern B is a **rectangle** and pattern C is a **line**.

1. The number 9 is a square number which is odd. Try to arrange 9 counters in a rectangle.

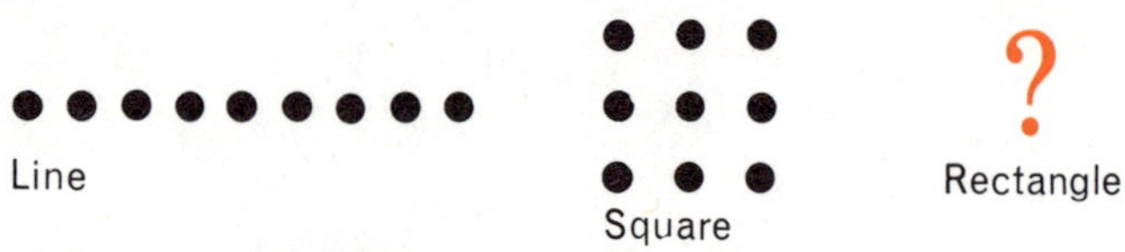

The number 18 can be shown as:

2. Can 18 be shown as a square **array of counters**?

   You will remember that {squares} is a subset of {rectangles}.
   A square is really a rectangle. Why?

3. Only one of these numbers can be shown as a square array. Two can be shown as rectangular arrays. Which one is square and which ones are rectangular? {3, 4, 5, 6, 7}

- **Rectangular numbers** can be shown as rectangular arrays.
- If the array is square then the number is **rectangular**.

Let's Try

1. Here are 36 coloured squares arranged in a rectangular array.

   Cut out 36 coloured squares each 2 cm by 2 cm and arrange them in as many rectangular patterns as possible.

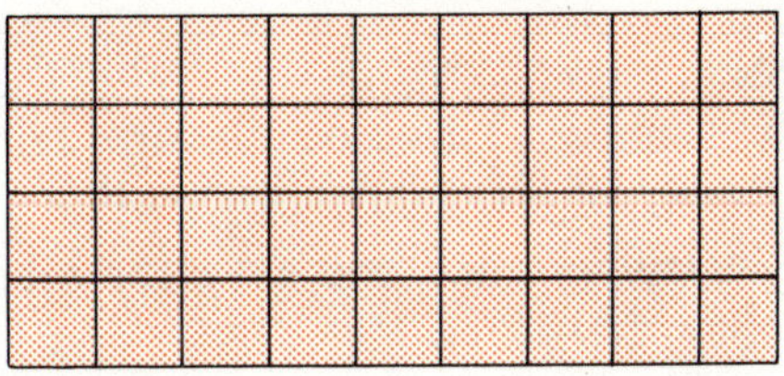

2. Draw all the possible arrays for each of these numbers:
   12; 18; 25; 28; 29; 30; 49.

3. Copy and complete this table.

Number	9	10	11	12	13	14	15	16	17	18	19	20
Square		×										
Rectangular	√	√										
Different arrays (Squares and rectangles)	1	2										

For example 10 is not square ×, Is rectangular √, and can be represented by:

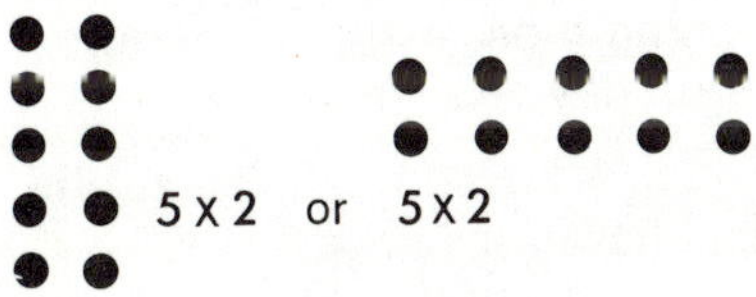

4. From {4, 9, 16, 25, 36, 49} find the square numbers which cannot be shown as a rectangular array.

5. Is it true that any square number **which is odd** can be arranged as a line or a square, but not as a rectangle? Try the first ten square numbers.

6. From the set of natural numbers {1, 2, 3, . . .} less than 30 write:
   a. the subset P of square numbers.
   b. the subset Q of rectangular numbers. Is P a subset of Q?
   c. the subset R of square numbers which are **odd**.
   d. the subset S of square numbers which are **even**.

7. a. Which of these numbers is **not** rectangular? 2, 4, 6, 8, 10
   b. Without drawing arrays, why can you be sure that 44, 102, 256, 1002 are rectangular numbers?

Let's Discover **More Tiling Patterns**

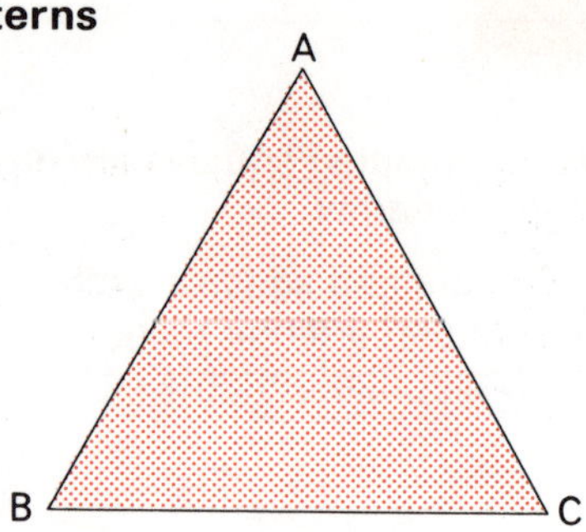

This is a special triangle.
The triangle is called **equilateral** because it has all its sides equal in length.

1. Draw, or trace, and cut out a triangle like ABC.
2. Tear off one corner of your triangle and see if it fits the other two corners.

- An equilateral triangle has three equal angles.

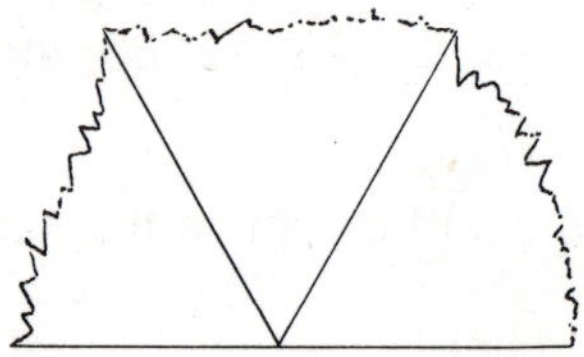

3. Tear off the other two angles of your triangle. Your three angles will fit together like this.

- The three angles of an equilateral triangle add up to 2 right angles.

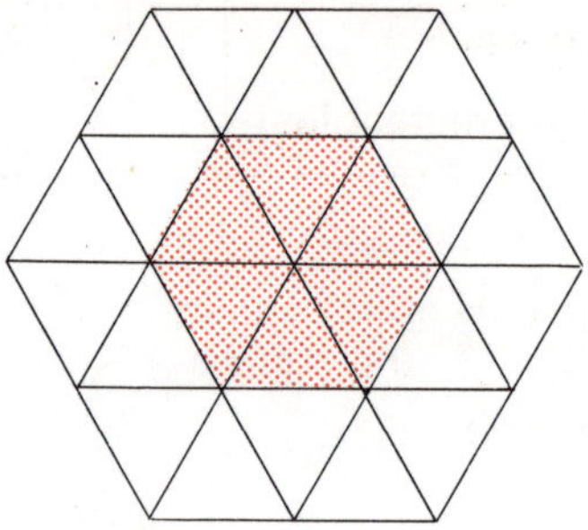

4. Cut out an equilateral triangle and see if you can build up a tessellation like this:

   The shaded hexagon is made up of 6 equilateral triangles.
   By adding more triangles a larger hexagon has been made.

5. Is the large hexagon similar to the shaded hexagon?
6. Build an even larger shape which is similar to the shaded hexagon.
7. These two shapes have been made from equilateral triangles.
   Use more triangles to make the next larger shape which is similar to triangle A.
   Make the next larger shape which is similar to the rhombus B.

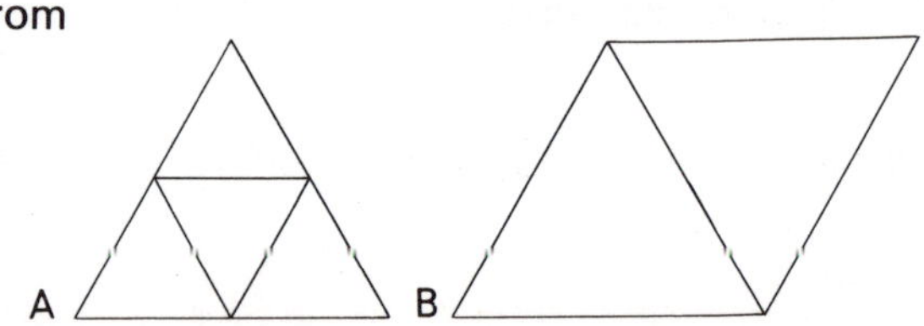

- These are shapes which have been made up from equilateral triangles.

Here are two more shapes which have been made up from equilateral triangles.

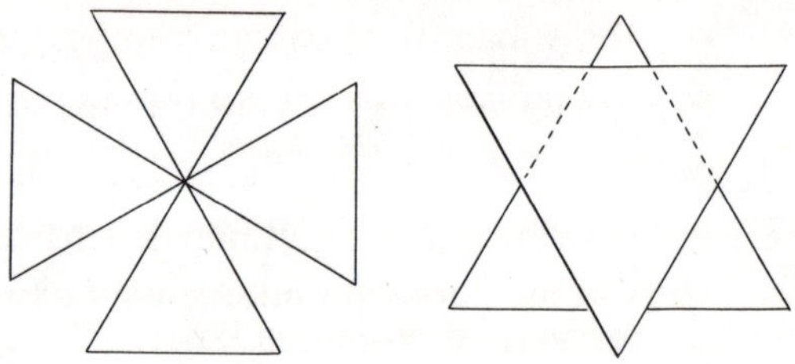

8. Try to tessellate shapes like these.

   Some shapes made up from equilateral triangles will not tessellate.

Let's Try

1. Build some more regular shapes with equilateral triangles.

2. a. Cut out 6 equilateral triangles of the same size. Join some or all of them together to make a shape which will tessellate.
   This is an example of a shape made from four equilateral triangles.
   b. Build up a tessellation with your shape.

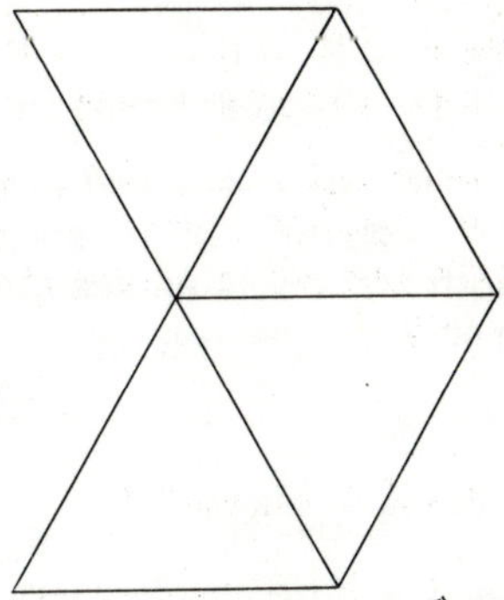

3. Repeat question **1** with other shapes.

4. a. Draw and cut out a triangle which is not equilateral.
   b. Will your triangle tessellate?
   Will any triangle tessellate?

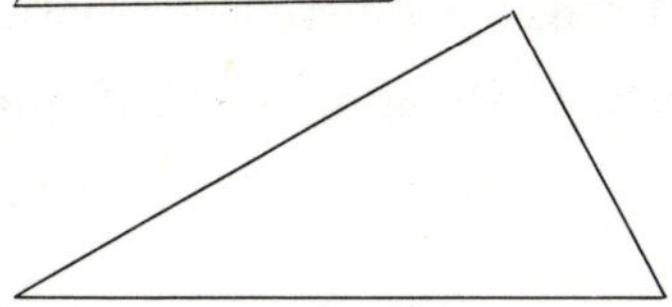

5. From equilateral triangles build up a trapezium and then tessellate your trapezium.

6. a. Draw and cut out an irregular quadrilateral.
   b. Will your quadrilateral tessellate?
   c. Will any quadrilateral tessellate?

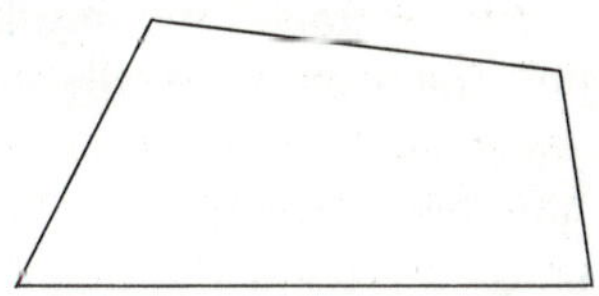

7. Here is a hexagon made from six equilateral triangles. Cuts may only be made on the black lines.
   a. How many rhombuses can it be cut into?
   Draw your own hexagon and make the cuts.
   b. In how many ways can you dissect the hexagon into rhombuses?
   c. How many trapeziums can you see in the diagram?
   d. Name all the shapes which can be cut from the diagram?

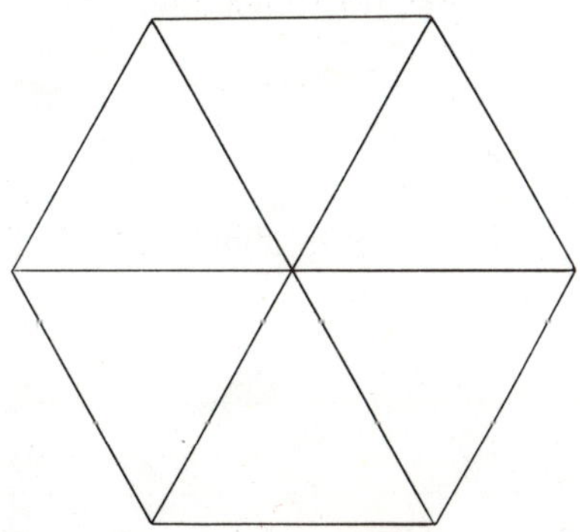

8. This picture shows a tiling pattern which has been started with octagons.
   What other shape besides the octagon do I need to finish tiling the floor?
   Build up a tessellation showing the finished tile pattern.

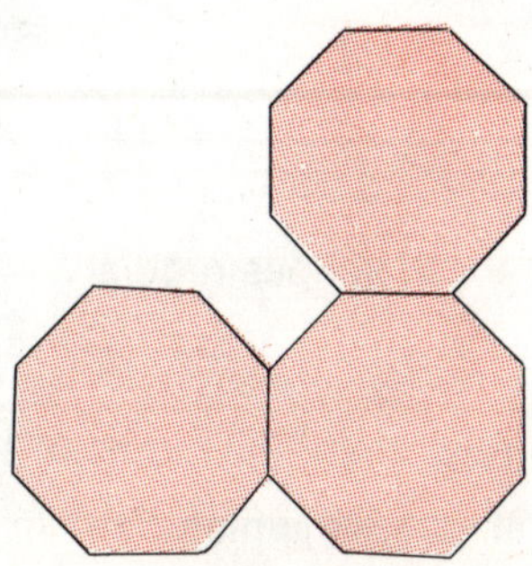

Let's Discover **Regular Shapes**

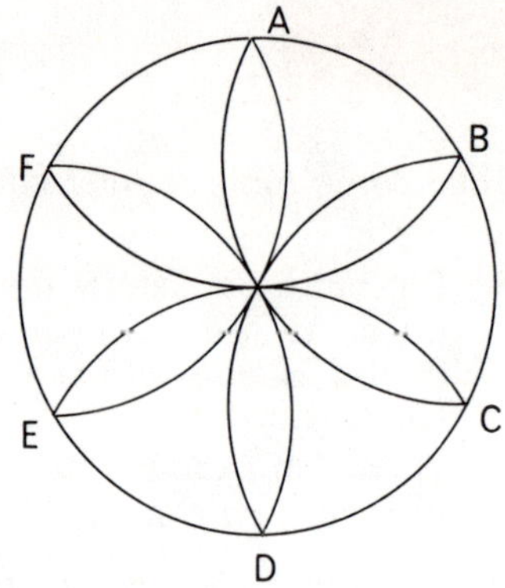

Copy this diagram which can be drawn using a pair of compasses.
By joining A → B → C → D → E → F → A you will draw a regular hexagon.

Any six-sided figure is called a hexagon. A shape is **regular** if all its sides are the same length and all its angles are the same size.

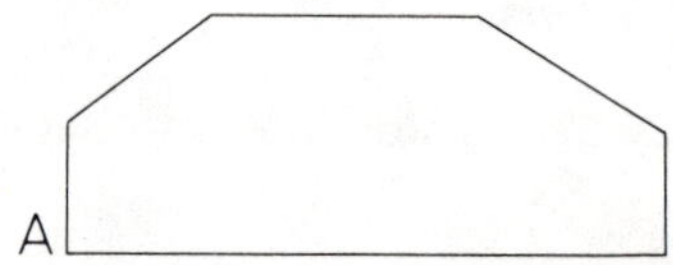

1. Is shape A a hexagon?
2. Is shape A a regular hexagon?
3. Which of the following shapes are regular?

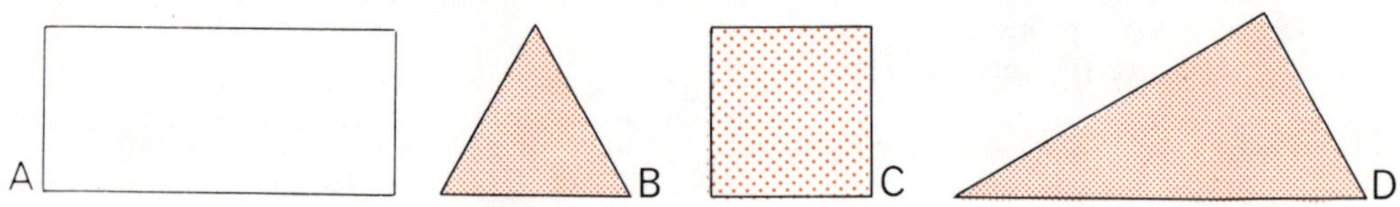

4. a. Draw, or trace, this regular hexagon and build up a tessellation with it.
   b. How many lines of symmetry like L_1 has this hexagon?
   c. How many lines of symmetry like L_2 has this hexagon?

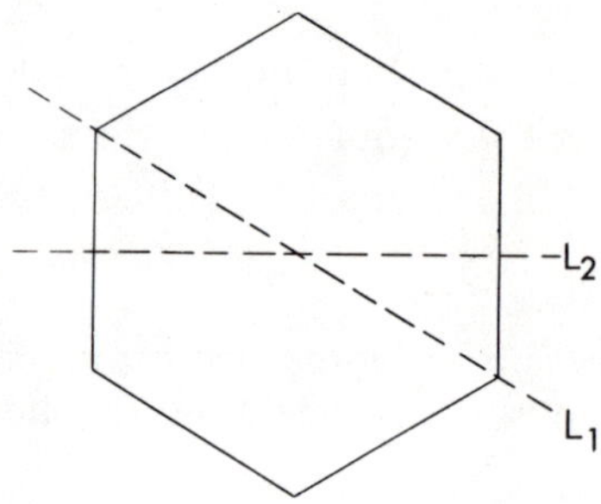

5. Here are two shapes which have lines of symmetry.

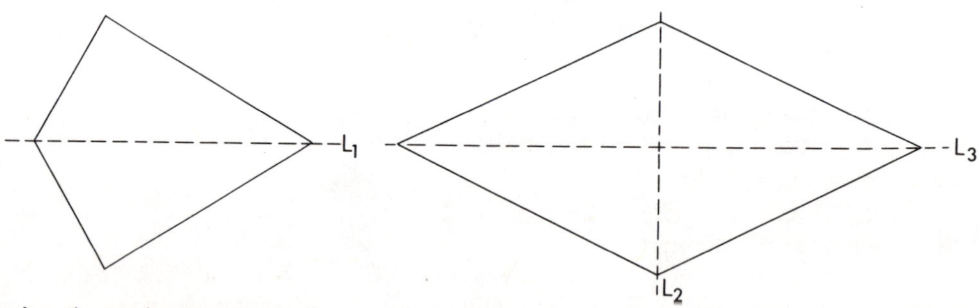

Are these shapes regular?

- A regular shape has lines of symmetry.
- Not all shapes which have lines of symmetry are regular.

6. Cut out a 10-cm square piece of paper. Fold it so that the creases of the paper look like the lines — — — — — —

Use the point where the creases cross as the centre and draw a circle with radius about 4 cm.

By joining A → B → C → D → E → F → G → H → A you will draw an **octagon**.

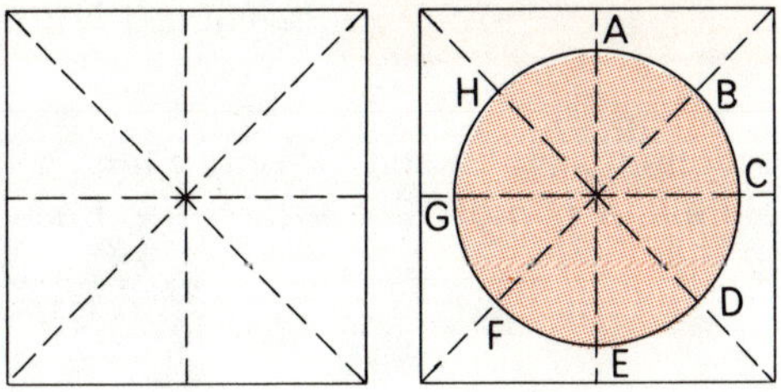

a. Is your octagon regular? b. How many lines of symmetry has it?

Let's Try

1. Which of the following shapes are:

a. regular b. quadrilaterals

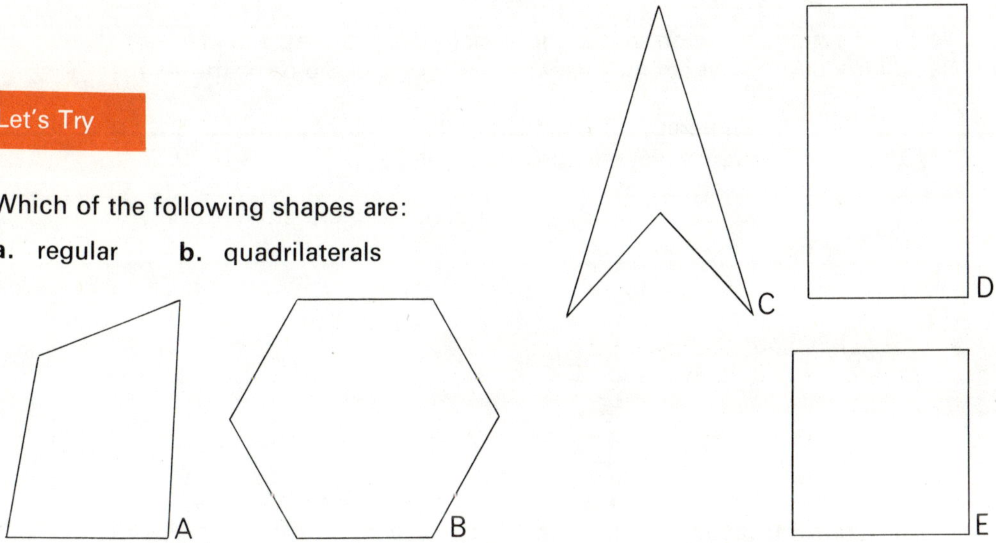

2. Draw the shape which you think is a regular triangle. Write a description of the triangle.

3. Draw the shape which you think is a regular five-sided figure. Write a description of the shape.

4. Trace, or draw, and cut out these two regular shapes. Build up some tiling patterns with them.

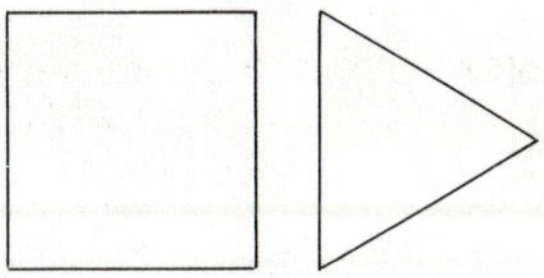

5. Trace, or draw, these three shapes.

a. Are all the shapes regular?

b. Find out which of these shapes will tessellate.

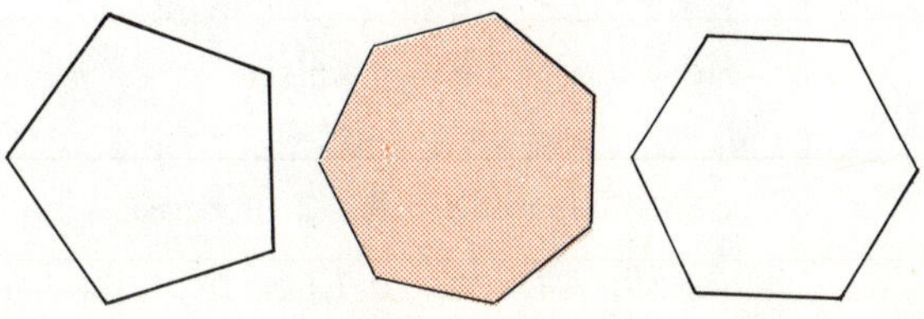

Let's Discover **Mixed Numbers**

Fractions can be shown on a number line.
Here is a number line which has been divided into tenths.

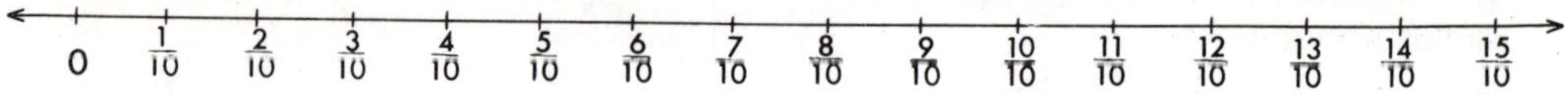

1. Use the number line to help you write these fractions in order of size with the smallest first.

$\frac{13}{10}$; $\frac{9}{10}$; $\frac{10}{10}$

We call the number which tells us how many parts the **numerator**.
We call the number which tells us what each part is the **denominator**.

2. $\boxed{\frac{13}{10}}$ The numerator of this fraction is . . .
The denominator of this fraction is . . .

3. I cut a cake into 10 parts. If I eat 10 parts, is there any cake left?
$\frac{10}{10}$ of a cake = 1 cake

- A fraction with the same numerator and denominator is renamed 1.

$\boxed{\frac{n}{n} = 1}$

$\frac{9}{10}$ **numerator** / **denominator**

This fraction is less than 1.
Remember that $\frac{10}{10} = 1$

4. $\{\frac{8}{10}, \frac{17}{10}, \frac{4}{5}, \frac{5}{5}, \frac{4}{4}\}$

a. Which two of these fractions are less than 1?

b. Which two of these fractions are equal to 1?

c. Will the numerator of a fraction which is less than 1 be greater than, equal to, or less than the denominator?

- A fraction which is less than 1 is called a **proper fraction**.

5. If I have 13 bars of chocolate to share amongst 10 children:

a. How many whole bars can they have each?

b. How many bars are left?

10)13
1 rem. 3

I can cut each extra bar into 10 pieces.

c. Can each child have 3 of these small pieces?

Each child will receive $1 + \frac{3}{10}$ of a bar of chocolate.

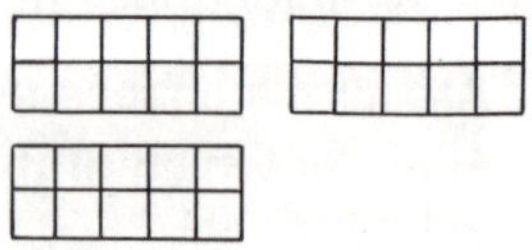

Instead of $10\overline{)13}$ we can write $10\overline{)13}$

1 rem. 3 $1\frac{3}{10}$

- $1\frac{3}{10}$ is called a **mixed number** because it is the sum of a whole number and a fraction.

The 13 children received $1\frac{3}{10}$ bars of chocolate.

6. a. Does $1\frac{3}{10} = 1+\frac{3}{10}$?   b. Does $1+\frac{3}{10} = \frac{10}{10}+\frac{3}{10} = \frac{13}{10}$?

- Any fraction with the numerator greater than the denominator is called an **improper fraction**.

7. Check this working which shows you how to change an improper fraction to a mixed number.

$\frac{13}{10} = \frac{10+3}{10} = \frac{10}{10}+\frac{3}{10} = 1+\frac{3}{10} = 1\frac{3}{10}$

Let's Try

1. Find the replacement for □ in:

a. $\frac{3}{10}+\frac{6}{10}=\frac{\square}{10}$ b. $\frac{3}{10}+\frac{7}{10}=\frac{\square}{10}$ c. $\frac{7}{10}+\frac{7}{10}=\frac{\square}{10}$

2. Which elements of $\{\frac{1}{5}, 1\frac{1}{2}, \frac{3}{2}, 3\frac{1}{2}, \frac{2}{10}, \frac{2}{7}\}$ are:

a. proper fractions   b. improper fractions   c. less than 1

d. mixed numbers   e. greater than 1   f. equal to 1

3. Write the answers to these divisions as mixed numbers:

a. $4\overline{)13}$ b. $5\overline{)36}$ c. $7\overline{)44}$ d. $9\overline{)80}$

e. $\frac{211}{7}$ f. $\frac{311}{10}$ g. $\frac{245}{6}$ h. $\frac{577}{8}$

4. Change these improper fractions to mixed numbers:

a. $\frac{17}{10}$ b. $\frac{13}{5}$ c. $\frac{103}{100}$ d. $\frac{67}{20}$ e. $\frac{44}{15}$

5. a. I have collected 7 pieces of wire each $\frac{1}{2}$ m long. If I join them end to end what is the total length in metres?

b. If it takes a boy $\frac{1}{4}$ hour to walk round a playground, how long should it take him to walk round 9 times?

Let's Discover

Different Terms

You have already discovered that many fractions can represent the same number.
For example: $\frac{1}{2} = \frac{2}{4} = \frac{3}{6} = \frac{5}{10}$

- Fractions which represent the same number are called **equivalent fractions**.

1. a. Does $\frac{1}{2} = \frac{1 \times 2}{2 \times 2} = \frac{2}{4}$? b. Does $\frac{1}{2} = \frac{1 \times 3}{2 \times 3} = \frac{3}{6}$?

   c. Does $\frac{1}{2} = \frac{1 \times 5}{2 \times 5} = \frac{5}{10}$? d. Does $\frac{3}{5} = \frac{3 \times 2}{5 \times 2} = \frac{6}{10}$?

   We multiplied the numerator and denominator of each fraction by the same number and made an equivalent fraction.

   $\boxed{\frac{3}{5}}$ is said to be in its **lowest terms** because there is no equivalent fraction with a smaller numerator and denominator.

   $\boxed{\frac{6}{10}}$ is in higher terms than $\boxed{\frac{3}{5}}$

   To change a fraction to **higher terms** without changing its value, **multiply** the numerator and denominator by the same number.

2. Instead of multiplying let's try dividing. Complete the following:

   a. $\frac{6}{10} = \frac{6 \div 2}{10 \div 2} = \ldots$ b. $\frac{2}{4} = \frac{2 \div 2}{4 \div 2} = \ldots$

   c. $\frac{10}{100} = \frac{10 \div 10}{100 \div 10} = \ldots$ d. $\frac{10}{100} = \frac{10 \div 5}{100 \div 5} = \ldots$

   e. Is $\frac{3}{5}$ in lower terms than $\frac{6}{10}$? f. Is $\frac{1}{2}$ in lower terms than $\frac{2}{4}$?

   g. Is $\frac{1}{10}$ in lower terms than $\frac{10}{100}$? h. Is $\frac{1}{10}$ in lower terms than $\frac{2}{20}$?

   To change a fraction to **lower terms** without changing its value, **divide** the numerator and denominator by the same number.

   Here is a new way of setting out your working for changing a fraction to lower terms:

   $\frac{10}{20}$ 2 will divide both 10 and 20.
   2 is a **common factor** of 10 and 20, so $\boxed{\frac{\cancel{10}^{5}}{\cancel{20}_{10}} = \frac{5}{10}}$

   $\frac{10}{20}$ 10 is a common factor of 10 and 20, so $\boxed{\frac{\cancel{10}^{1}}{\cancel{20}_{2}} = \frac{1}{2}}$

- This new method of setting out is called **cancelling**.

Let's Try

1. What number do we multiply the numerator and denominator by to change the first fraction to the second?

   a. $\frac{1}{2} = \frac{5}{10}$ b. $\frac{3}{4} = \frac{75}{100}$ c. $\frac{2}{5} = \frac{40}{100}$ d. $\frac{3}{5} = \frac{21}{35}$

2. What number do we divide the numerator and denominator by to change the first fraction to the second?

   a. $\frac{14}{21} = \frac{2}{3}$ b. $\frac{55}{100} = \frac{11}{20}$ c. $\frac{52}{100} = \frac{13}{25}$ d. $\frac{70}{100} = \frac{14}{20}$

3. Find the replacement for each △ in:

   a. $\frac{15}{30} = \frac{\triangle}{6}$ b. $\frac{28}{100} = \frac{\triangle}{25}$ c. $\frac{\triangle}{100} = \frac{3}{4}$

4. The factors of the numerator and the denominator of $\frac{8}{20}$ are shown in orange.

   $8 \leftarrow \{1, 2, 4, 8\} = A$
   $20 \leftarrow \{1, 2, 4, 5, 10, 20\} = B$

   a. Which factors are common to set A and set B? $A \cap B = \ldots$
   b. Change $\frac{8}{20}$ to its lowest terms.

5. a. Fill in all the factors for sets A and B and find $A \cap B$.

   $30 \leftarrow \{\ldots\ldots\ldots\ldots\} = A$ $A \cap B$
   $100 \leftarrow \{\ldots\ldots\ldots\ldots\} = B$ $= \{\ldots\ldots\}$

   b. Change $\frac{30}{100}$ to its lowest terms.

6. Use the cancelling method to change these fractions to their lowest terms:

   a. $\frac{20}{40}$ b. $\frac{25}{50}$ c. $\frac{15}{50}$ d. $\frac{12}{100}$ e. $\frac{14}{100}$

7. Change these fractions to their lowest terms and then to mixed numbers:

   a. $\frac{14}{4}$ b. $\frac{22}{20}$ c. $\frac{22}{10}$ d. $\frac{32}{20}$ e. $\frac{530}{100}$

8. Tom said that his tie is $\frac{1}{4}$m long. Bill said, "No it isn't, it is $\frac{25}{100}$m long." Is $\frac{25}{100}$m the same length as $\frac{1}{4}$m?

9. Is $1\frac{1}{2}$kg the same as $\frac{1500}{100}$kg?

10. Which is the greater volume: $1\frac{3}{4}$ litres or $\frac{170}{100}$ litre?

Let's Discover **Relations**

Here is a sketch map of an island.
The point (3, 2) is marked with x.
A point in a plane can be described by an ordered pair.

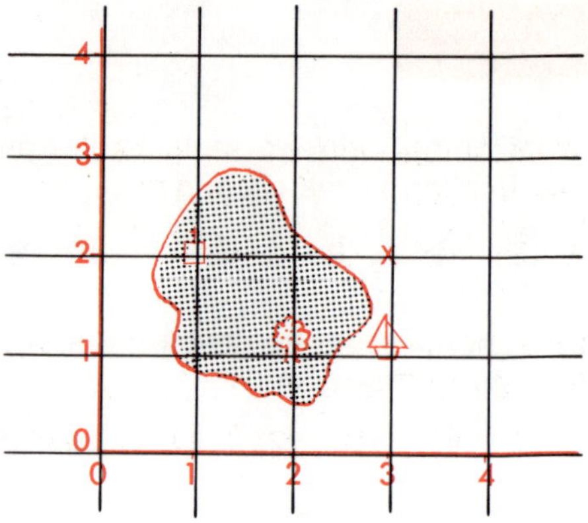

1. a. What would you find at (1, 2)?
   b. What would you find at (2, 1)?
   c. Where is the yacht?

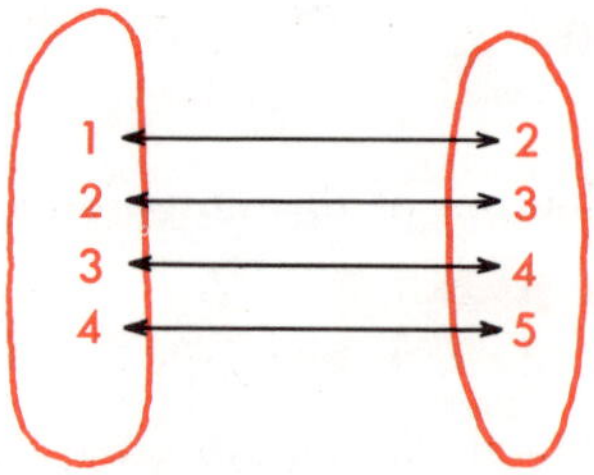

2. a. Look for the relationship between the numbers which have been mapped on this diagram.

   This is a list of the pairs.

   {(1, 2), (2, 3), (3, 4), (4, 5)} = A

   b. If the set A had contained more elements could it have contained: (5, 6), (6, 7), (8, 7)?

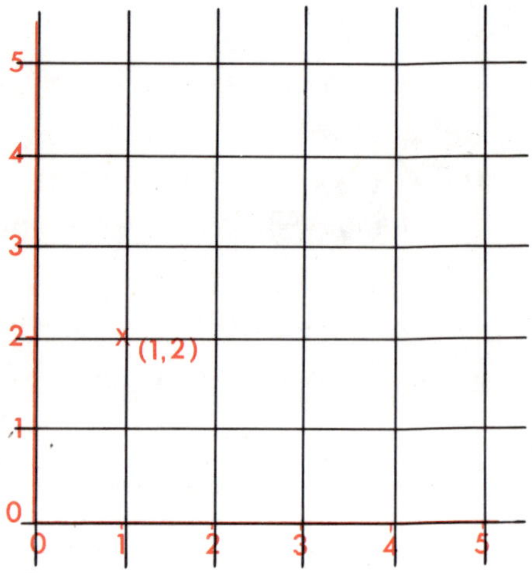

3. a. Copy a pair of axes like these onto squared paper.
   b. Represent each ordered pair from the set A as a point on your graph. The point (1, 2) has already been graphed for you.
   c. What do you notice about the pattern of points?

   You have drawn a graph of your set of ordered pairs.

- A set of ordered pairs is called a **relation**.

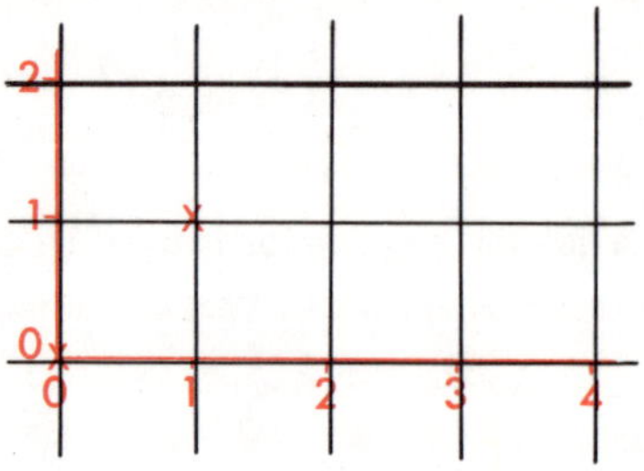

4. Here is another relation.

   {(0, 0), (1, 1), (4, 2), (3, 9), (4, 8)} = C

   The graph of relation C has been started for you.
   Draw a graph of the relation for yourself.

Let's Try

1. a. The ordered pair which gives the position of Tom's nose is (. . . , . . .).
   b. What part of Tom is at (2, 3)?
   c. Is Tom's ear at (3, 2)?
   d. Where is the tip of Tom's beard?
   e. What would happen to Tom if I made a horizontal cut from (2, 1) to (3, 1)?

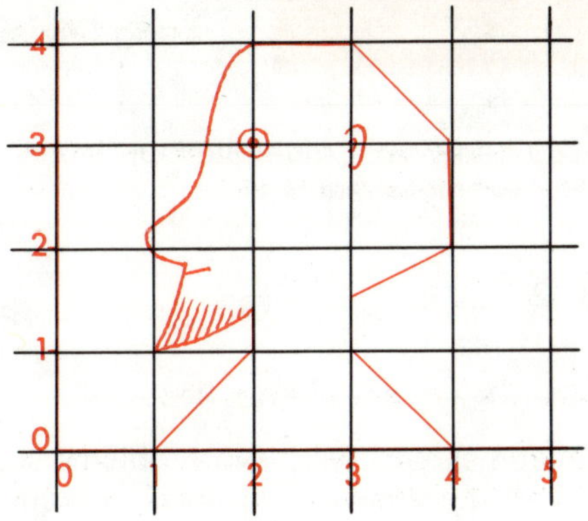

2. Here is a message in code:
   (1, 3), (3, 2), (2, 3), (1, 1), (1, 2),
   (2, 1), (1, 2), (2, 2), (3, 3), (3, 1)
   a. Write the letter represented by each ordered pair.
   b. Is the message true?

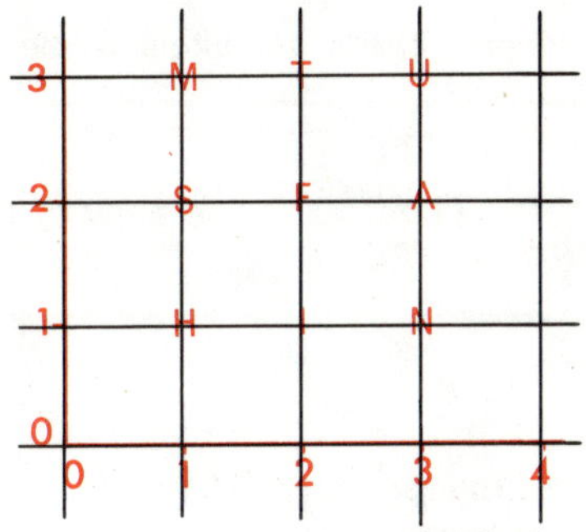

3. a. Write the set of ordered pairs which is shown on this graph.
   b. What do we call a set of ordered pairs?
   c. Copy this graph and join up the points to form a square.
   d. Write the ordered pair which represents one point inside the region you have enclosed.
   e. Draw all the lines of symmetry on your square.

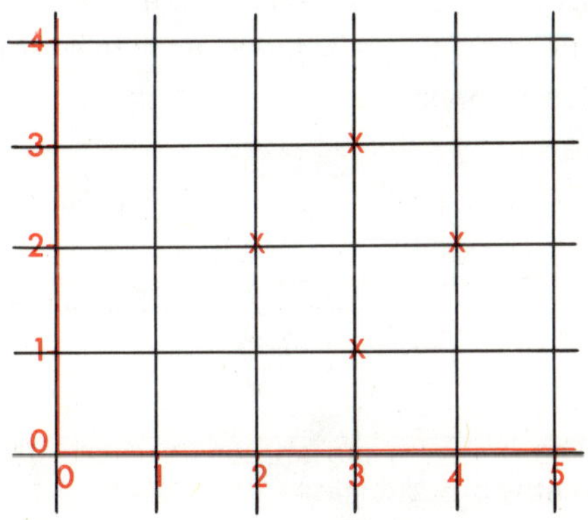

4. The ordered pairs in each of the following relations have a pattern. Fill in the missing pair and graph the relation.
   a. {(5, 0), (5, 1), (5, 2), (. . . , . . .), (5, 4), (5, 5)}
   b. {(0, 4), (. . . , . . .), (2, 4), (3, 4), (4, 4), (5, 4)}
   c. {(0, 0), (1, 2), (2, 4), (. . . , . . .), (4, 8), (5, 10)}

Let's Discover **Long Division**

If you know your multiplication tables, you will be able to find the replacement in many examples such as:

$8 \div 2 = \triangle$

Since $4 \times 2 = 8$ then $8 \div 2 = 4$

Sometimes you will have problems such as $89 \div 21$
To find the answer you need to work like this:

$21\overline{)89}$ with ? on top. How many 2s in 8? $2\overline{)8}$ with ? on top.

Since $2 \times 4 = 8$, there are likely to be **4 twenty-ones** in **89**.

Let's check:

```
          4
      21)89
21×4 = 84 ⟶ 84
          5
```

The answer is 4 rem. 5 or $4\frac{5}{21}$

Here is another problem: $121 \div 21$.
Working in the same way:

$21\overline{)121}$ with ? on top. How many 2s in 12? $2\overline{)12}$ with ? on top.

Since $2 \times 6 = 12$, there are likely to be **6 twenty-ones** in **121**.

Let's check:

```
           21)121
21×6 = 126 ⟶
```

21×6 is too big so let's try 21×5

```
                5
           21)121
21×5 = 105 ⟶ 105
               16
```

The answer is 5 rem. 16 or $5\frac{16}{21}$

The answers are shown in these questions. Check them by using your new method of division.

a. $87 \div 23 = 3\frac{18}{23}$ **b.** $131 \div 21 = 6\frac{5}{21}$

Let's use our new method on $486 \div 23$
$48 \div 2 = 24$

Since $48 \div 2 > 9$ we:

divide the hundreds and tens,

```
              2
         23)486
23×2 = 46 ⟶ 46
              2
```

then bring down the 6 and divide again.

```
             21
         23)486
             46
             26
23×1 = 23 ⟶ 23
              3
```

The **answer** is 21 rem. 3

Let's Try

1. Copy and complete these multiplication tables:

a.

×	2	—	—	8
7	14	28	42	—

b.

×	—	—	—	7
9	9	27	45	—

$7 \times 2 = 14$ and $14 \div 7 = 2$

2. The first division has been started for you. Finish it off and then use the same methods for parts **b**, **c**, **d**, and **e**.

a. $152 \div 25 = \ldots$ rem. $\ldots$

$$\begin{array}{r} 6 \\ 25\overline{)152} \end{array}$$

$25 \times 6 = \ldots \longrightarrow$

b. $160 \div 32$ **c.** $159 \div 32$

d. $168 \div 24$ **e.** $172 \div 24$

3. Are the answers to these divisions **greater than 10**, **less than 10**, **or equal to 10**?

a. $250 \div 25$ **b.** $251 \div 25$ **c.** $249 \div 25$

d. $392 \div 37$ **e.** $299 \div 31$ **f.** $272 \div 19$

g. Work out the answer to each division.

4. My car will travel 12 km on 1 litre of petrol. How much petrol should I need for a journey of 180 km?

5. The length of Charlie Stroller's pace is 47 cm. How many paces will he take to cover a distance of:

a. 470 cm **b.** 4·7 m **c.** 470 m

6. When Jill was 10 it cost her father £9·90 to provide a party for 22 children. What was the cost per child?

7. A 2-kg bobbin of wool has to be divided into hanks weighing 50g. How many hanks can be made from the bobbin of wool?

8. $24 \times 14 = 336$ Use this to help you write the answers to:

a. $336 \div 14$ **b.** $336 \div 24$ **c.** $336 \div 12$

9. I thought of a number and divided it by 25. My answer is 6 remainder 5. What was the number I thought of?

10. Divide 1776 by 24

Let's Discover **A Use for Graphs**

Suppose that you are a shopkeeper and you have to serve three customers with wire from rolls which cost 4p, 6p and 8p per metre.
Mr. ●, Mr. ■ and Mr. ▲ are your customers.

1. a. Copy this table and fill in the missing amounts.

Price of Wire	Mr. ● Length	Mr. ● Cost	Mr. ■ Length	Mr. ■ Cost	Mr. ▲ Length	Mr. ▲ Cost
4p per metre 6p per metre 8p per metre	3 m 2 m 1 m	12p 12p 8p	7 m 10 m 6 m		6 m 20 m 8 m	
Total cost for each customer		32p		£1·36		£2·08

b. How many calculations did the shopkeeper make in order to complete this table?

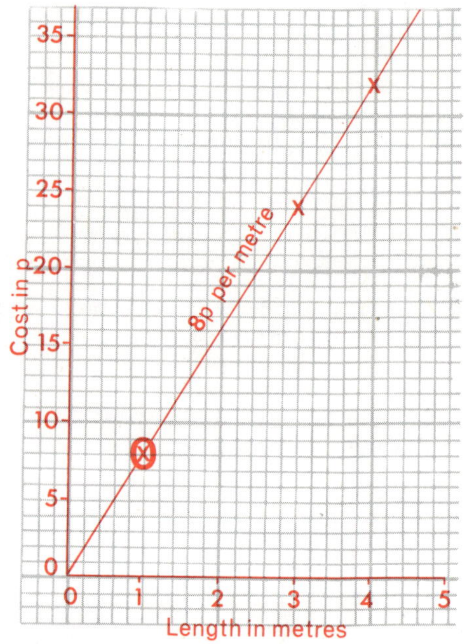

Our shopkeeper could have saved a lot of calculations by drawing graphs like this. One has been started for you.

The ordered pair (1, 8) has been marked with ⓧ on the graph.

2. a. Write the two ordered pairs which have been marked with x.

b. Three points have been marked on the graph.
Do we really need three points?
Wouldn't two be enough?

c. Copy this graph but extend the axes to 8 metres and 64p.

d. From your graph find the cost of:
4 metres at 8p per metre — 4·5 metres at 8p per metre
6 metres at 8p per metre — 6·5 metres at 8p per metre

e. Check your answers to **d** by calculation.

If the shopkeeper charged me 12p for a length of the 8p wire I would have to read across → and down ↓ to find that

12p is the cost of $1\frac{1}{2}$ metres.

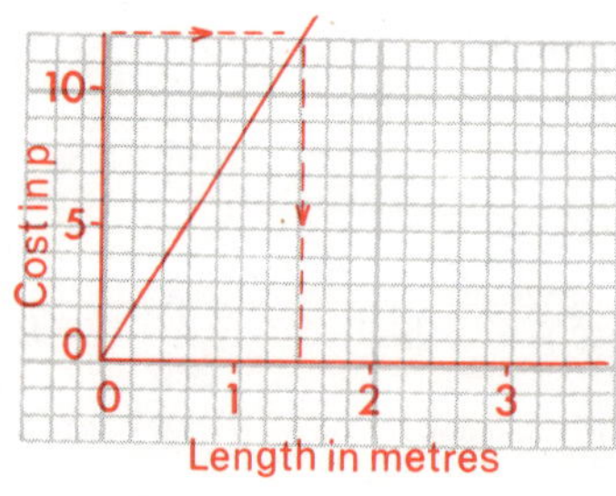

Let's Try

1. Here is part of a graph showing the cost of bottles of lemonade.

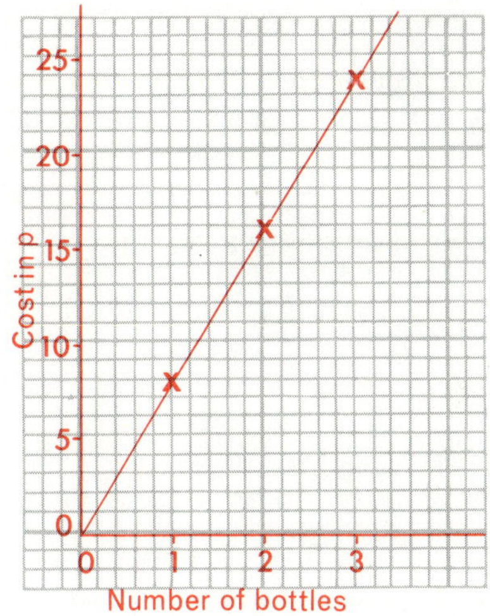

   a. What is the cost of 3 bottles?

   b. What is the cost of 1 bottle?

   c. Complete this set of ordered pairs which is represented by points on the graph.
   {(1, . . .), (. . . , 16), (3, . . .)}

   d. What is the cost of 10 bottles?

   e. Draw the graph showing the cost of bottles of lemonade but extend your axes to 10 bottles and 80p.

   f. From your graph find the cost of 6, 9, and 8 bottles.
   Check your answers.

   g. **Use your graph** to find out how many bottles you could buy for 24p, 40p, 32p, and 56p.

2. Jane is cycling steadily and travels 8 km in 1 hour. Draw your horizontal axis 0 to 16 km and your vertical axis 0 to 120 minutes.

   a. Complete this set of ordered pairs. The first number in each pair is the distance travelled in km and the second number is the time taken in min.
   {(0, 0), (4, . . .), (8, . . .), (. . . , 90)}

   b. Mark points x on your graph which represent the ordered pairs. Draw a line through your four points.

   c. **From your graph** find out how long it will take Jane to cycle:
   16 km, 10 km, 2 km, 6 km, 14 km.

   d. **From your graph** find out how far Jane will cycle in:
   45 min, $\frac{1}{4}$ hour, $1\frac{1}{2}$ hours, 75 min, 7·5 min.

   e. Check your answers to **c** and **d** by calculation.

3. Tom's mother asked him to peel some potatoes. It took him 40 minutes to peel 24 potatoes. Do you think that Tom was slow?
Assume that each potato takes the same time and draw a graph showing "Time taken to peel potatoes". Use your graph to find out how many potatoes he peels in 30 min; 15 min; 5 min; 10 min.

Let's Discover — Another Set of Numbers

On page 8 you learned about factors.

1. Are there any factors of:

a. 2 **b.** 3 **c.** 5 **d.** 7
e. 11

Look at the number 8 or

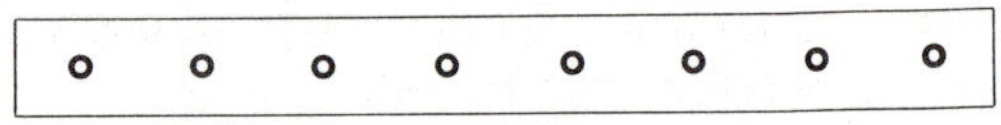

8 is **divisible** by 2

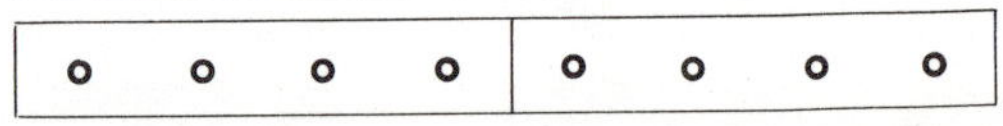

and by 4

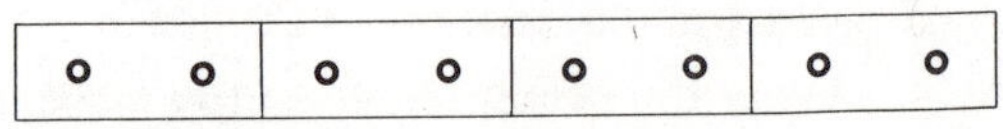

8 can be arranged as a rectangular array.

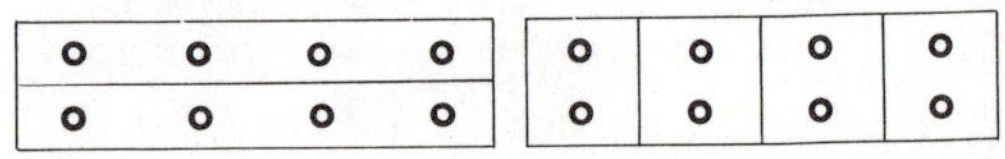

Look at the number 7.
7 has no factors other than 1 and itself and cannot be arranged in a rectangular array.

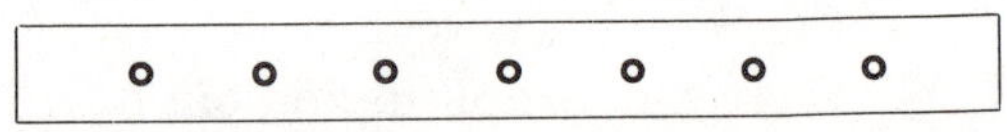

● A number which has no factors other than **1** and itself is called **prime**.

7 is divisible only by 1 and 7.
7 cannot be arranged as a rectangular array.
7 is a **prime number**.

8 is divisible by 1, 2, 4, 8.
8 can be arranged as a rectangular array.
8 is **not a prime number**.

2. Look back at question **1**. Check that 2, 3, 5, 7 and 11 are prime number.

3. a. Here is part of a number square. Copy and complete the square up to 100.

b. Shade all the multiples of 2 but not 2.
After 2 the next unshaded number is 3, so shade all multiples of 3 in a different colour.
The next unshaded number after 3 will be 5, so shade all multiples of 5 in a third colour.
Continue with this process and then make a list of your unshaded numbers.
You have sieved the set of prime numbers less than 101. You must take out the number 1.

c. Look at your set of prime numbers.

● 2 is the only **even prime** number.

Let's Try

1. Copy each strip as many times as necessary to show the different ways in which each strip can be divided to form rectangular arrays.

   a. 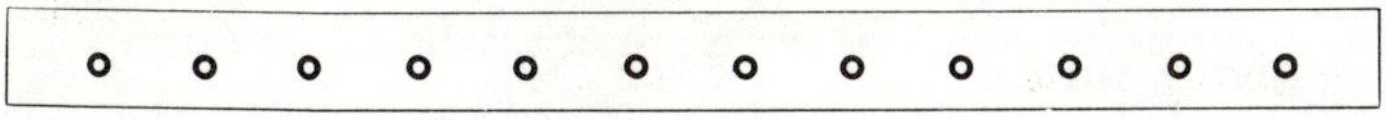 (12)

   b. 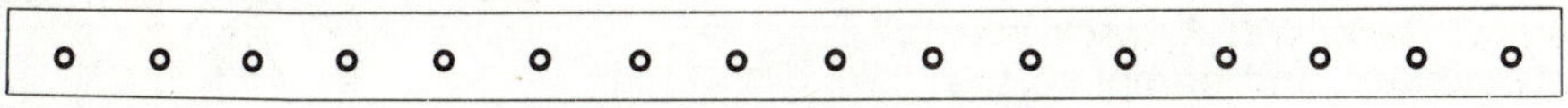 (16)

2. By referring to your sieve complete the following:

   a. R is the set of prime numbers less than 25. R = {2, 3, . . . , 23}
   b. S is the set of prime numbers less than 50. S = {2, 3, . . . , 47}
   c. The number of primes less than 25, n(R) = . . .
   d. The number of primes less than 50, n(S) = . . .

3. Find the smallest and the largest prime numbers which are:

   a. between 50 and 120   b. between 90 and 140

4. Find all the prime numbers by which each of the following is divisible:

   a. 6   b. 9   c. 10   d. 12   e. 14   f. 15

5. What is the smallest natural number which is divisible by both:

   a. 2 and 3   b. 3 and 5   c. 2 and 5
   d. 4 and 2   e. 4 and 6   f. 6 and 9

6. 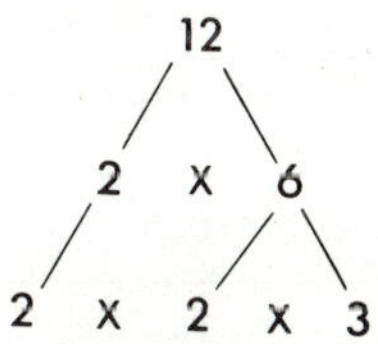

   Here is a factor tree for 12.
   Are the factors on the bottom row, 2 and 3, prime numbers?
   Are there any other factors of 12 which are prime?

7. Write factor trees for:

   a. 24   b. 16   c. 18

   From each factor tree write the set of factors which are prime.

8. a. Are all these fractions in their lowest terms?

   $\frac{3}{10}$ $\frac{4}{5}$ $\frac{2}{9}$

   b. Complete the following sentence.
   Either the numerator or the denominator of each fraction is a . . . number.

Let's Discover — More Lengths

John said to his teacher,
"I live five minutes away from school."

It is sometimes useful to give a **time** instead of a **distance**.

1. Find out how long it takes you to get to school, and complete the following sentence.
"I live . . . minutes away from school."

We need to know how John gets to school. Does he fly, or cycle, or walk, or . . . ?

John walks to school and he wants to know how far it is.
He can't walk very far in 5 minutes, so he could count his paces.

In the playground a friend helped John to find the distance covered by 10 of his paces.

2. At normal walking pace John found that 10 paces covered about 8·5 metres, so
1 pace measures about
$\frac{8 \cdot 5}{10}$ m = . . . m = . . . cm

3. Complete the following for John.
I take about 610 paces between home and school, so it is about 610×0·85 m

$$\begin{aligned} &= (600 \times 0{\cdot}85) + (\ .\ .\ .\ \times 0{\cdot}85)\text{ m} \\ &= (6 \times 85) + 8{\cdot}5\text{ m} \\ &= .\ .\ .\text{ m} \end{aligned}$$

4. **a.** Make a column graph of "The length of our paces" for 10 children in the class.

b. From your graph which child has the longest pace and which child has the shortest pace?

c. From your graph which child will take the most paces and which child will take the least paces in walking 100 m?

Some distances are not easily measured in paces. Would you like to try measuring the height of a horse in paces?

This is a picture of a pony which is 12 **hands** high.

1 **hand** = 10 cm

so the pony is 12×10 cm
= 120 cm high.

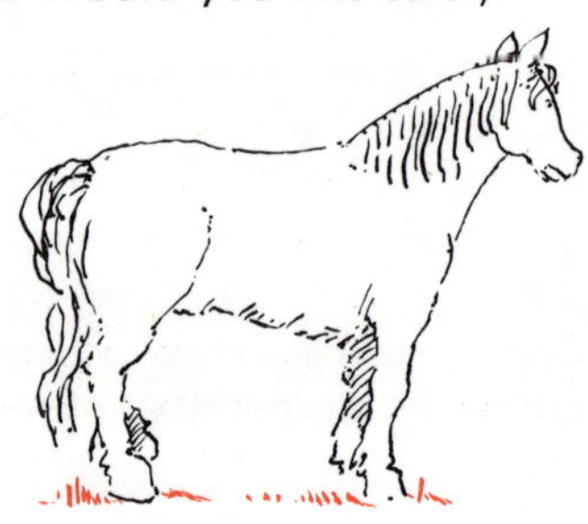

5. Carefully draw round your own hand and measure its width across your palm. Call this measure **my hand**.

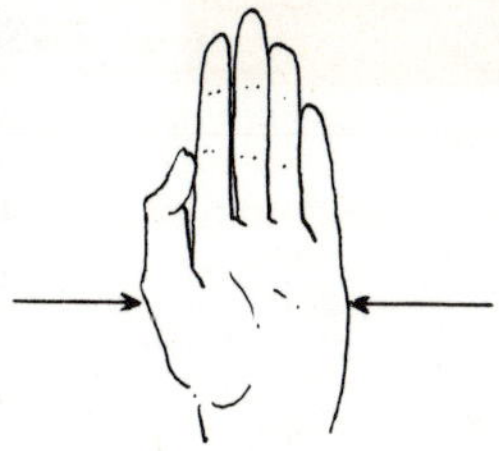

- Remember to use your hand and pace to help you make measurements.

Let's Try

1. A hand is 10 cm. How tall are these things in cm?

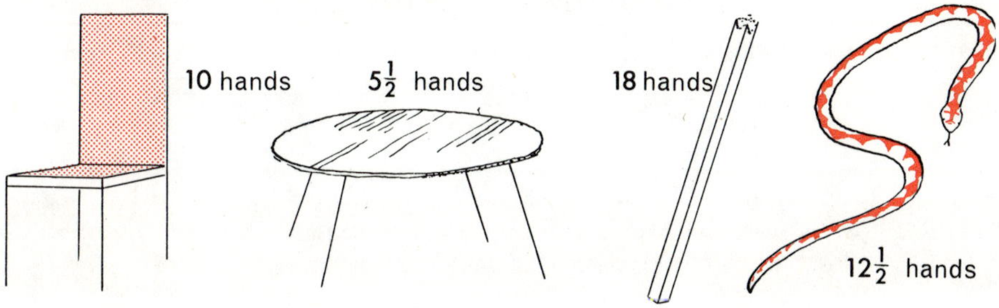

2. If your pace is 60 cm, how many metres have you covered if you walked the following number of paces?

   a. 10 b. 5 c. 100 d. 1000

   How many paces would it take to walk:

   e. 600 cm f. 6 m g. 6·5 m

   h. 60 m

3. Here is a picture of a park. The numbers show the paces taken by a boy whose pace is 50 cm.

   a. How wide (in metres) is the park?

   b. How long (in metres) is the park?

   c. Write the perimeter of the park in paces and then in metres.

   d. How many metres do you think it is from the fir tree to the flagpole?

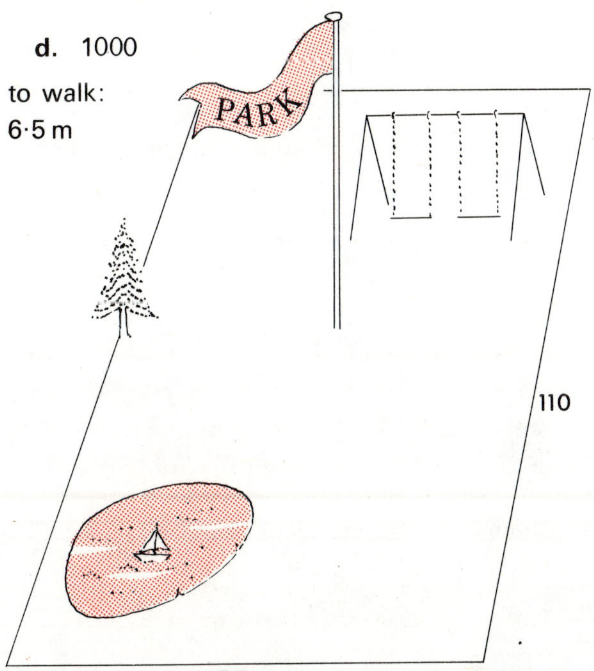

4. Draw a plan of your classroom. Show measurements in your paces.

Let's Discover

Addition and Subtraction of Fractions

Complete these three examples:

1. $\frac{1}{4}+\frac{3}{4}=\frac{?}{4}=1$

2. $\frac{3}{10}+\frac{4}{10}=\frac{?}{10}$

3. $\frac{2}{5}-\frac{1}{5}=\frac{?}{5}$

These three examples will remind you of the way in which we add or subtract fractions which have **the same denominator**.

4. Change the terms of: **a.** $\frac{2}{5}=\frac{?}{10}$ **b.** $\frac{1}{2}=\frac{?}{10}$

This number line is marked off in tenths.

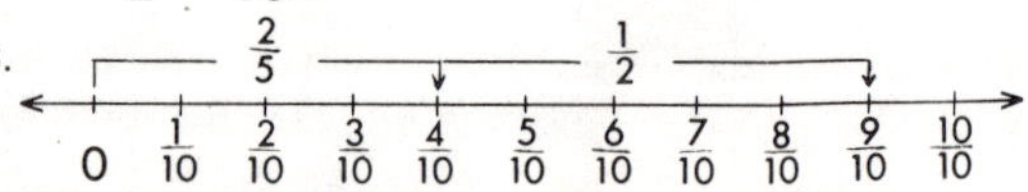

5. Check that $\frac{2}{5}+\frac{1}{2}$ or $\frac{4}{10}+\frac{5}{10}=\frac{9}{10}$

- When adding fractions which have different denominators we need to change the terms so that the fractions have the same denominator.

6. Complete these examples and check on the tenths number line.

a. $\frac{3}{5}+\frac{1}{10}=\frac{?}{10}+\frac{1}{10}=\frac{?}{10}$ **b.** $\frac{2}{5}+\frac{1}{10}=\frac{?}{10}+\frac{1}{10}=\frac{?}{10}=\frac{1}{2}$

7. **a.** In question **5**, is 10 a multiple of 5 **and** 2?

b. In question **6**, is 10 a multiple of 5 **and** 10?

The new denominator must be a multiple of **both** the old denominators.

To find $\frac{1}{4}+\frac{1}{5}$ we need to consider multiples of 4 and 5.

A = {Multiples of 4}
= {4, 8, 12, 16, 20, 24, 28, 32, 36, 40, . . .}
B = {Multiples of 5}
= {5, 10, 15, 20, 25, 30, 35, 40, 45, . . .}

8. Are 20 and 40 common to both set A and set B? A ∩ B = { . . . , . . . }
so 20 is the **lowest common multiple** of 4 and 5.

9. Change $\frac{1}{4}$ and $\frac{1}{5}$ to twentieths. $\frac{1}{4}=\frac{?}{20}$ and $\frac{1}{5}=\frac{?}{20}$

$\frac{1}{4}+\frac{1}{5}=\frac{5}{20}+\frac{4}{20}=\frac{9}{20}$

This number line is marked off in tenths.

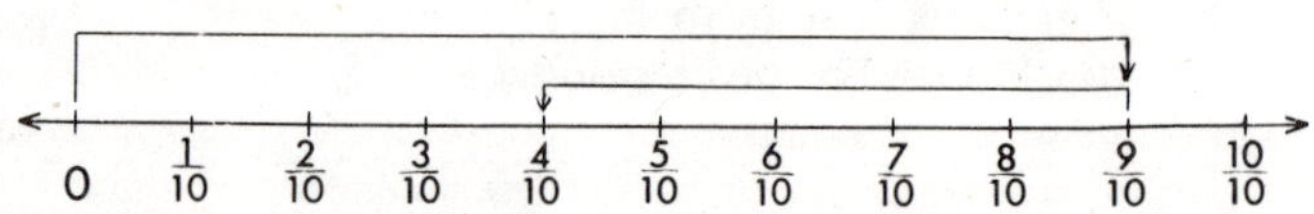

10. Check on the number line that $\frac{9}{10}-\frac{1}{2}$ or $\frac{9}{10}-\frac{5}{10}=\frac{4}{10}$

- When subtracting fractions which have different denominators we need to change the terms so that the fractions have the same denominator.

11. Complete these examples and check on the tenths number line.

a. $\frac{3}{5}-\frac{3}{10}=\frac{?}{10}-\frac{3}{10}=\frac{?}{10}$ **b.** $\frac{4}{5}-\frac{1}{2}=\frac{?}{10}-\frac{?}{10}=\frac{?}{10}$

Let's Try

1. This number line is marked off in tenths.

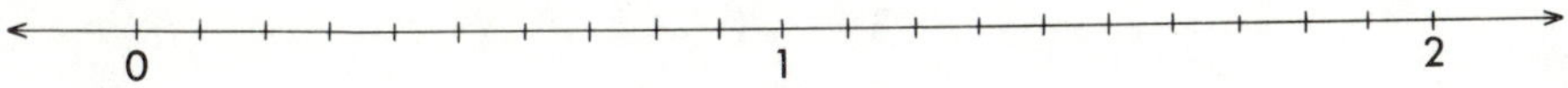

Draw a similar number line for each of these examples and find:

a. $\frac{3}{10}+\frac{3}{5}$ **b.** $\frac{3}{5}-\frac{3}{10}$ **c.** $\frac{1}{2}+\frac{7}{10}$ **d.** $\frac{7}{10}-\frac{1}{2}$ **e.** $1\frac{1}{2}-\frac{7}{10}$

2. What is the lowest number which has the following as factors?

a. 2 and 5 **b.** 5 and 10 **c.** 2 and 10 **d.** 4 and 10
e. 100 and 10

3. Complete these additions:

a. $\frac{1}{4}+\frac{3}{10}$ **b.** $\frac{3}{5}+\frac{3}{10}$ **c.** $\frac{7}{20}+\frac{2}{5}$ **d.** $\frac{31}{100}+\frac{1}{10}$ **e.** $\frac{21}{50}+\frac{3}{100}$

4. Find the answer to:

a. $\frac{3}{10}-\frac{1}{5}$ **b.** $\frac{3}{4}-\frac{7}{10}$ **c.** $\frac{7}{20}-\frac{1}{10}$ **d.** $\frac{31}{100}-\frac{7}{50}$ **e.** $\frac{1}{4}-\frac{1}{10}$

5. Add the two fractions and then **without working out** write the answer to the subtractions.

$\frac{1}{2}+\frac{3}{10}=\ldots$ $\frac{4}{5}-\frac{1}{2}=\ldots$ $\frac{4}{5}-\frac{3}{10}=\ldots$

6. John's stick of rock is $\frac{3}{4}$m long. Mary's is only $\frac{3}{10}$m long.
As a fraction of a metre:

a. What is the total length?
b. What is the difference in length?

7. Mary and Jane each baked a cake.
As a fraction of a kg:

a. What is the total weight of both cakes?
b. What is the difference in weight?

Let's Discover **Multiplication of Fractions**

1. Find the replacement for each △.

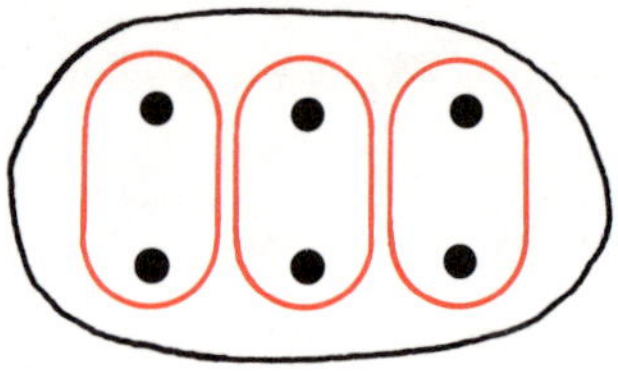

a. In a set of 6 there are △ sets of 2.
b. $\triangle \times 2 = 6$
c. $6 \div \triangle = 2$
d. $\frac{1}{3}$ of $\triangle = 2$

2. Draw a number line like this one.

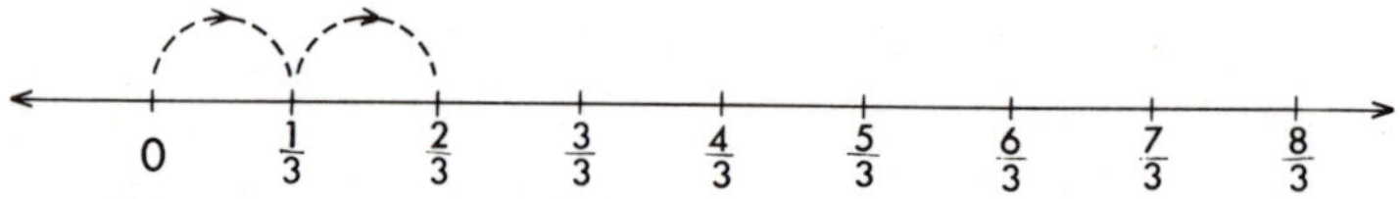

a. Start at 0 and show 6 steps of $\frac{1}{3}$ on your number line.

Is it true that:

b. $\frac{6}{3} = 2$ **c.** 6 of $\frac{1}{3} = 2$

- The word "of" between fractions means multiply.

d. Use the number line to find | 7 of $\frac{1}{3}$ | or | $7 \times \frac{1}{3}$ |

Use this region as a **unit**.
$\frac{3}{4}$ of the region has been shaded.

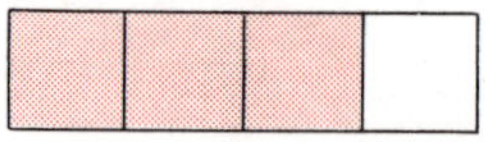

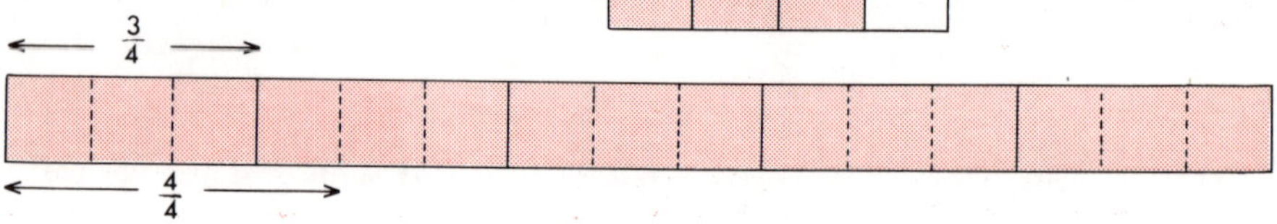

3. **a.** How many pieces each $\frac{3}{4}$ of a unit have been placed end to end?
b. How many whole units could you cut from the long piece?
c. How many **quarters** would be left?
d. Is it true that $4 \times \frac{3}{4}$ unit $= \frac{3}{4}+\frac{3}{4}+\frac{3}{4}+\frac{3}{4} = \frac{12}{4}$?
e. Fill in the missing number. $4 \times \frac{3}{4} = \frac{? \times 3}{4} = \frac{12}{4} = 3$
f. Look at the long strip again. Is $\frac{3}{4}$ of $4 = 3$?
g. Fill in the missing number. $\frac{3}{4} \times 4 = \frac{3 \times ?}{4} = \frac{12}{4} = 3$

- When multiplying a whole number by a fraction, or a fraction by a whole number, multiply the numerator of the fraction by the whole number.

Let's Try

1. Copy each number line and show:

   a. $3\times\frac{1}{2}$ (with •) b. $\frac{1}{3}$ of 2 (with •)

   c. $4\times\frac{1}{2}$ (with ×) d. $\frac{2}{3}$ of 3 (with ×)

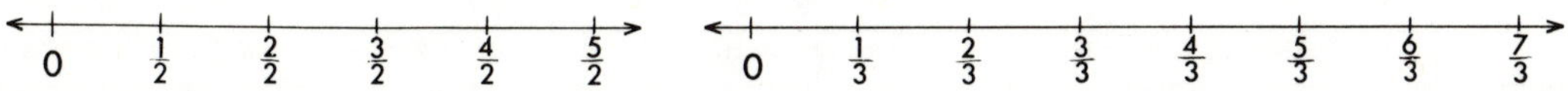

2. Find the answer to:

   a. $3\times\frac{3}{10} = \ldots$ b. $2\times\frac{2}{5} = \ldots$ c. $7\times\frac{1}{10} = \ldots$

   d. $\frac{3}{10}\times 2 = \ldots$ e. $\frac{1}{5}\times 4 = \ldots$ f. $\frac{3}{10}\times 3 = \ldots$

3. In these examples the answer is a mixed number.

   a. $3\times\frac{3}{4} = \ldots$ b. $3\times\frac{3}{5} = \ldots$ c. $\frac{2}{3}\times 7 = \ldots$

   d. $\frac{2}{5}\times 7 = \ldots$ e. $\frac{9}{10}\times 5 = \ldots$ f. $10\times\frac{31}{100} = \ldots$

4. Name the missing fraction in:

   a. $\ldots\times 3 = \frac{3}{4}$ b. $2\times\ldots = \frac{4}{5}$ c. $10\times\ldots = 9$ d. $3\times\ldots = \frac{21}{4}$

5. How many p are there in:

   a. $\frac{1}{4}$ of £1 b. $\frac{3}{4}$ of £1

6. A housewife took $\frac{3}{4}$ hour to dress a chicken. How long would it take her to dress 3 chickens?

7. $\frac{2}{7}$ of 3 weeks is . . . week, or . . . days.

8. My nail tin holds 500 nails. If I have used $\frac{2}{5}$ of the nails:

   a. how many nails have I used? b. how many nails are left?

9. Complete the following:

   a. In a $\frac{1}{2}$-kg tin of coffee there are $\frac{1}{2}\times 1000$ g = . . . g of coffee.

   b. In a tin holding $\frac{1}{5}$ kg there are . . . g of coffee.

   c. In a tin holding $\frac{4}{5}$ kg there are . . . g of coffee.

   d. $\frac{3}{4}$ of the perimeter of a square is 9 cm. The perimeter is . . . cm.

Let's Discover

Some Right Angles

Here is a set of shapes:

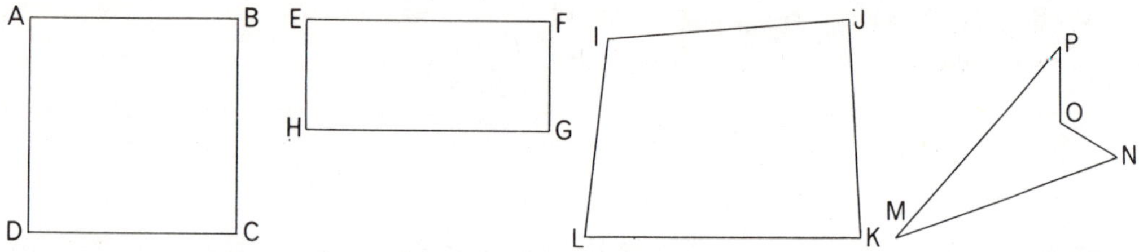

Each shape has four edges. The edges of shape ABCD are $\overline{AB}$, $\overline{BC}$, $\overline{CD}$ and $\overline{DA}$. Any closed shape which has four edges is called a **quadrilateral**. ABCD is a special quadrilateral which is called a square.

1. Test the angles of the square.
2. Measure the line segments: $\overline{AB}$, $\overline{BC}$, $\overline{CD}$ and $\overline{DA}$.

- A **square** has four edges of the same length and its angles are right angles.

Here is a shape which has four right angles.

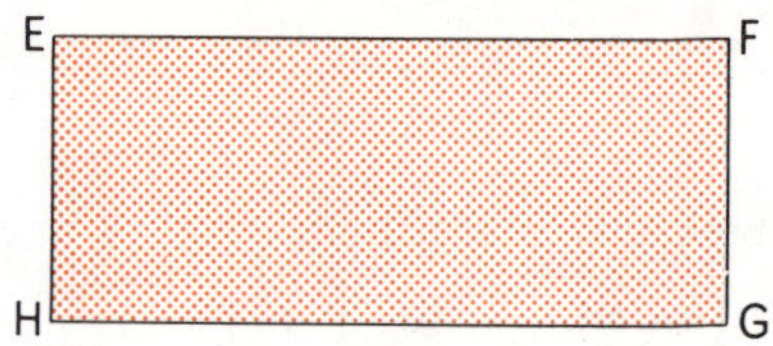

3. Test each angle.
4. Measure the line segments: $\overline{EF}$, $\overline{FG}$, $\overline{GH}$ and $\overline{HE}$.
5. Name two pairs of equal sides.

- A rectangle is a quadrilateral which has opposite edges equal in length and its angles are right angles.

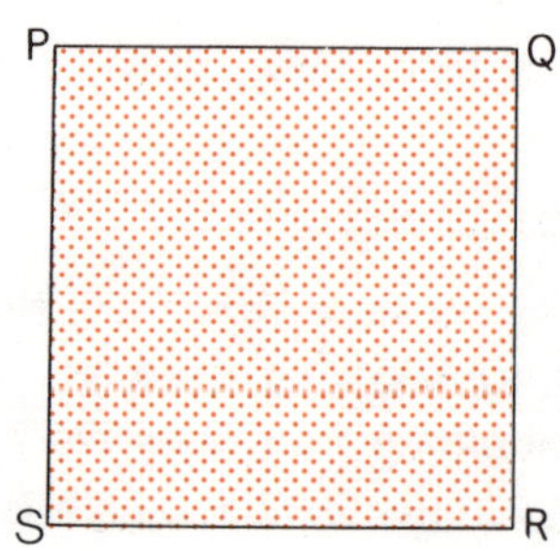

6. Test the angles of PQRS.
7. Is $m(\overline{PQ}) = m(\overline{SR})$?
8. Is $m(\overline{QR}) = m(\overline{PS})$?
   The shape PQRS could be called a rectangle because:

   PQRS has four right angles,

   $m(\overline{PQ}) = m(\overline{SR})$, and $m(\overline{QR}) = m(\overline{PS})$

   The shape PQRS is a special rectangle which we call a **square** because:

   $m(\overline{PQ}) = m(\overline{SR}) = m(\overline{QR}) = m(\overline{PS})$.

Let's Try

1. Here is a set of shapes:

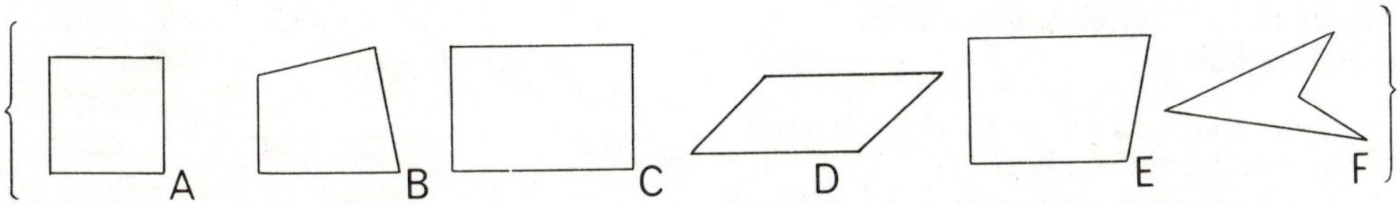

From this set:

   a. How many right angles can you see in each shape?
   b. Write the subset of shapes which are squares.
   c. Write the subset of shapes which are rectangles but not squares.
   d. Write the subset of shapes which are quadrilaterals.
   e. Write the subset of shapes which are not quadrilaterals.

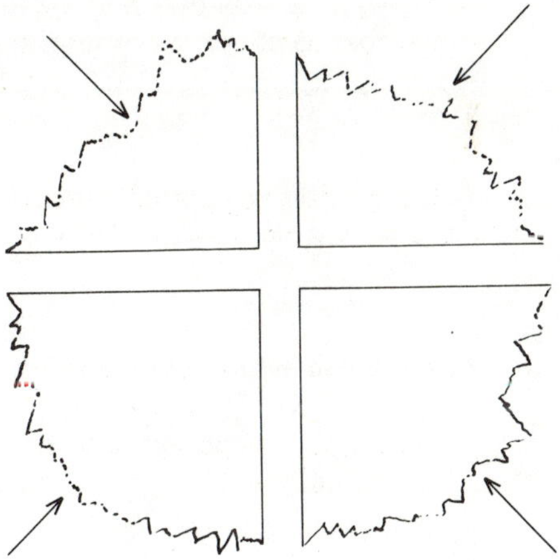

2. Draw and cut out a square. Tear off the four corners and fit them together. Complete this sentence. The angles of a square make . . . right angles.

3. Draw and cut out a rectangle. Tear off the four corners and fit them together. The angles of a rectangle make . . . right angles.

4. Draw and cut out any quadrilateral you like. Tear off the four corners and fit them together.

5. Trace, or draw, and cut out these shapes. Cut along the lines marked ---. Try to arrange your five pieces into a square. (The two crosses must be the same size.)

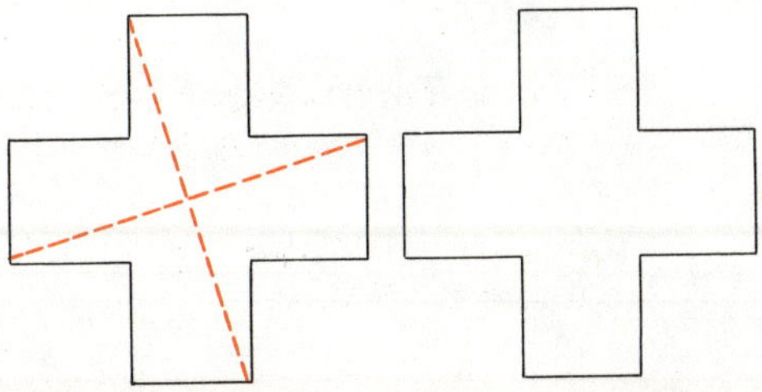

6. a. Name as many squares as possible from this figure.
   b. From this figure, name as many rectangles as possible which are not squares.

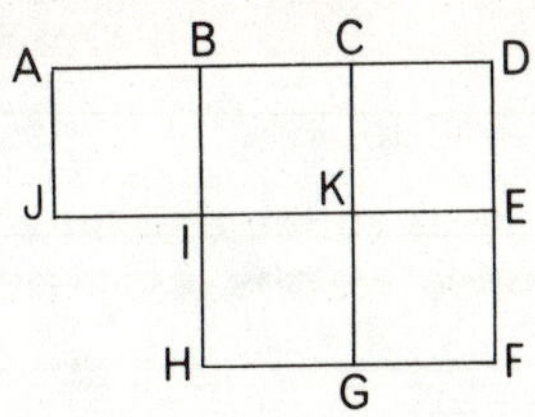

Let's Discover **Tare and Net**

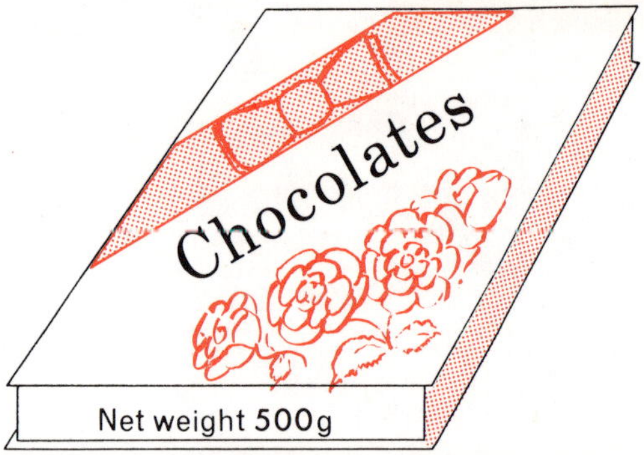

Make a collection of labels and cartons showing the weight of contents.
For example this chocolate box contains 500 g.

1. How many of these boxes together would contain 1 kg?
Look carefully at the chocolate box lid. Can you see that it says,
Net weight 500 g?

- **Net weight** is the weight of chocolates inside the box.

You wouldn't be very pleased if you bought a $\frac{1}{2}$-kg box of chocolates and found that $\frac{1}{4}$ kg of it was box and wrappings.
Remember to check whether the weight you buy is the **total** weight or the **net** weight.

2. If 500 g is the weight of chocolate and wrappings, what is the weight of chocolate if the wrappings weigh 120 g?

3. a. Find out where your nearest **public weigh bridge** is.
A weigh bridge is used for weighing vehicles.

After loading his truck with sand the driver uses the weigh bridge.
Altogether his truck and load weigh 2500 kg.

b. The weight of the truck is 1500 kg.
What weight of sand has been loaded?

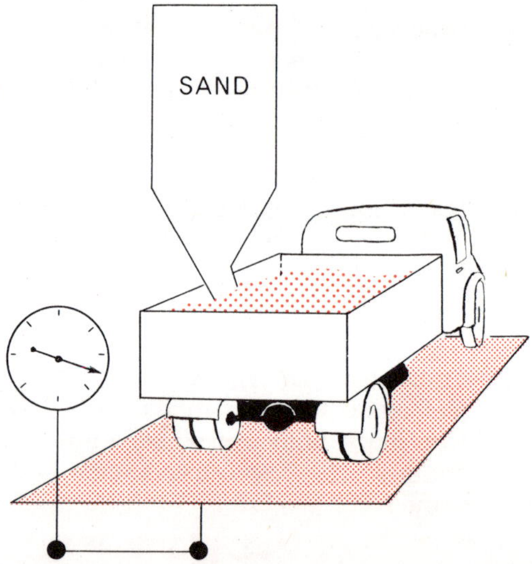

The words **Tare weight 1·5 tonne** will be shown somewhere on this truck.

1 tonne (t) = 1000 kg so
1·5 tonne = 1500 kg

- **Tare weight** is the weight of the unladen vehicle.

Be careful when looking at lorries. Make sure that it is safe.

Look at the jar of coffee and tin of beans at the top of the next page.
The jar of coffee weighs less than the tin of beans.

4. The jar of coffee costs 35p and the tin of beans costs 5p.

Complete this working:

a. 500 g of coffee cost 35p so 100 g of coffee cost $\frac{35}{5}$p = . . . p

b. 1000 g of beans cost 5p so 100 g of beans cost $\frac{5}{10}$p = . . . p

Let's Try

1. Change these tonne weights to kg.

a. 2·5 t b. 1·4 t c. 1·45 t d. 0·5 t

2. Change these weights to tonnes.

a. 2000 kg b. 2500 kg c. 1300 kg d. 600 kg

3. A British Rail car transporter has a tare weight 61·5 tonnes. How many kg is this?

4. Use a small jar (or tin) for this experiment.

a. Find the weight of the jar full of: dried peas; water; sand; teacher's chalk; . . .

b. Draw a column graph showing the weight of each jar full.

c. Find the weight of the empty jar and then find the net weight of each jar full.

5. Here is a tin of fruit which **weighs 500 g**.

a. What is the weight of 2 tins of fruit: in g; in kg?

b. Find the weight of 10 tins of fruit: in g; in kg?

c. How much fruit will there be in 10 tins of fruit?

d. Of a full tin:
What fraction of the weight is fruit?
What fraction of the weight is tin?

Let's Discover **How to Convert**

Tom said, "I have eaten $\frac{1}{2}$ of my sweets."
Bill said, "Tom has eaten 0·5 of his sweets."

1. If Tom had 10 sweets to start with and had eaten 5, were both Tom and Bill right?

2. Here is a set of **common fractions**: $A = \{\frac{1}{2}, \frac{7}{10}, \frac{3}{100}\}$
and a set of **decimal fractions**: $B = \{0{\cdot}7, 0{\cdot}03, 0{\cdot}5\}$
Which common fraction is equivalent to which decimal?

3. Fill in the missing numbers in:

a. $0{\cdot}13 = \frac{1}{10} + \frac{?}{100} = \frac{?}{100} + \frac{3}{100} = \frac{13}{100}$

How many hundredths are equal to 0·13?

b. $1{\cdot}14 = 1 + \frac{?}{10} + \frac{4}{100} = 1 + \frac{?}{100} + \frac{4}{100} = 1\frac{14}{100} = 1\frac{?}{50}$

How many hundredths are equal to 0·14?

c. $1{\cdot}23 = 1\frac{?}{100}$　　d. $1{\cdot}04 = 1\frac{?}{100} = 1\frac{?}{25}$

Here is a completed example showing how to convert a **decimal** to a **common fraction**.　$2{\cdot}18 = 2\frac{\not{1}\not{8}\,9}{\not{1}\not{0}\not{0}\,50} = 2\frac{9}{50}$

4. Follow through this working and write in the decimal answer.

$\frac{23}{100} = \frac{20}{100} + \frac{3}{100} = \frac{2}{10} + \frac{3}{100} = 0{\cdot}\ldots$

Here are some fractions which have been rewritten as decimals.

a. $\frac{3}{100} = 0{\cdot}03$　　b. $\frac{13}{100} = 0{\cdot}13$　　c. $1\frac{27}{100} = 1{\cdot}27$

5. In these examples the terms of the fractions are being changed.
Fill in the missing numbers.

a. $\frac{3}{5} = \frac{?}{10}$　so $\frac{3}{5} = 0{\cdot}\ldots$

b. $\frac{7}{50} = \frac{?}{100}$　so $\frac{7}{50} = 0{\cdot}\ldots$

c. $\frac{3}{25} = \frac{?}{50} = \frac{?}{100}$　so $\frac{3}{25} = 0{\cdot}\ldots$

- In order to change a fraction to a decimal we need to change the terms of the fraction so that the denominator is 10 or 100.

For example,

d. $\frac{3}{4} = \frac{15}{20} = \frac{75}{100} = 0{\cdot}75$　　e. $\frac{2}{5} = \frac{4}{10} = 0{\cdot}4$　　f. $\frac{3}{25} = \frac{6}{50} = \frac{12}{100} = 0{\cdot}12$

Let's Try

1. Change these common fractions to decimal fractions.

   a. $\frac{9}{100}$ b. $\frac{16}{100}$ c. $\frac{75}{100}$ d. $\frac{19}{10}$ e. $1\frac{75}{100}$ f. $\frac{125}{100}$

2. Find the replacement for each △ in:

   a. $0{\cdot}14 = \frac{\triangle}{100}$ b. $2{\cdot}04 = 2\frac{\triangle}{100}$ c. $10{\cdot}07 = 10\frac{\triangle}{100}$

3. Change these decimal fractions to common fractions in their lowest terms.

   a. 0·35 b. 10·06 c. 1·06 d. 2·45 e. 0·65 f. 10·13

4. Change these common fractions to decimal fractions.

   a. $\frac{11}{50}$ b. $\frac{3}{20}$ c. $\frac{4}{25}$ d. $\frac{17}{20}$ e. $\frac{49}{50}$ f. $\frac{24}{25}$

5. a. Change 0·27 m to a fraction of a metre.
   b. How many cm are the same length as 0·27 m?

6. Complete the following: $25\text{ cm} = \frac{?}{100}\text{ m} = 0\cdot\ .\ .\ .\text{ m}$

7. Write these lengths as a decimal of a metre and then change the decimal to a common fraction.

   a. 25 cm b. 45 cm c. 130 cm d. 224 cm

8. Write these weights as a decimal of a kg and then change the decimal to a common fraction.

   a. 600 g b. 750 g c. 250 g d. 2400 g

9. Here is a picture of a carpet.

   a. What fraction of the carpet is coloured orange?
   Write your fraction in its lowest terms.
   b. What fraction of the carpet is white?
   Write your answer in its lowest terms.
   c. Write the answer to **a** as a decimal.
   d. Write the answer to **b** as a decimal.

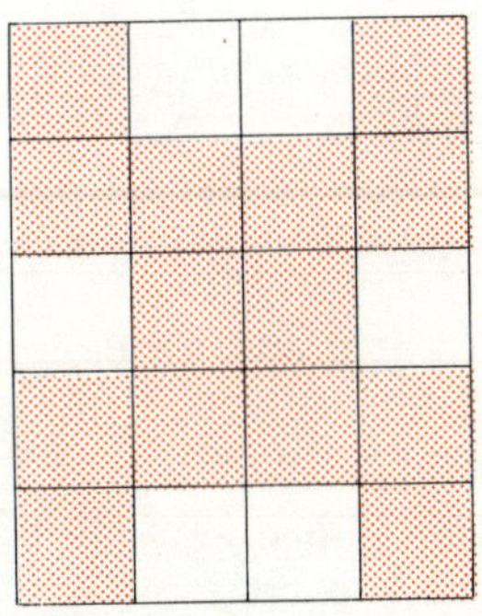

Let's Discover **More Angles**

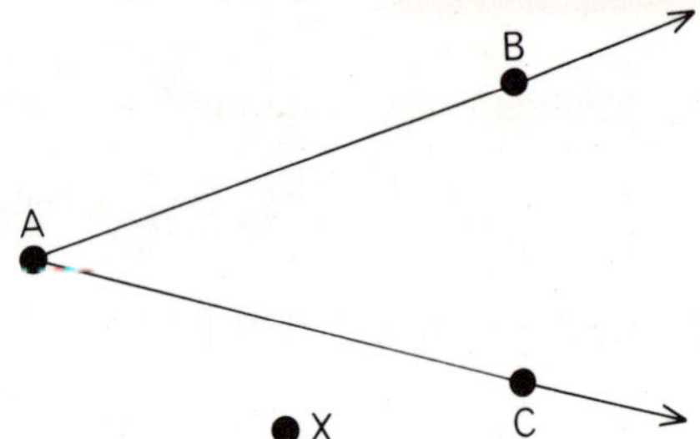

Angle BAC is made up by the union of rays $\overrightarrow{AB}$ and $\overrightarrow{AC}$.

The angle is named as $\angle$ CAB or $\angle$ BAC.

1. Name the point of intersection of the two rays.
2. Name each of the following angles in two different ways:

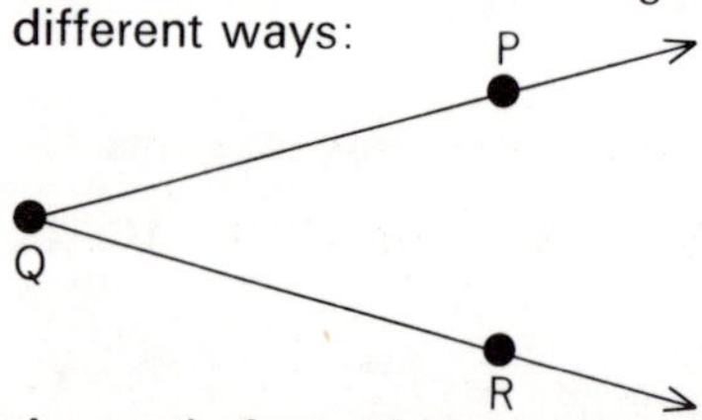

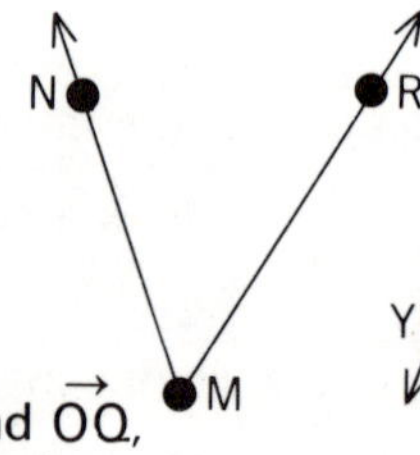

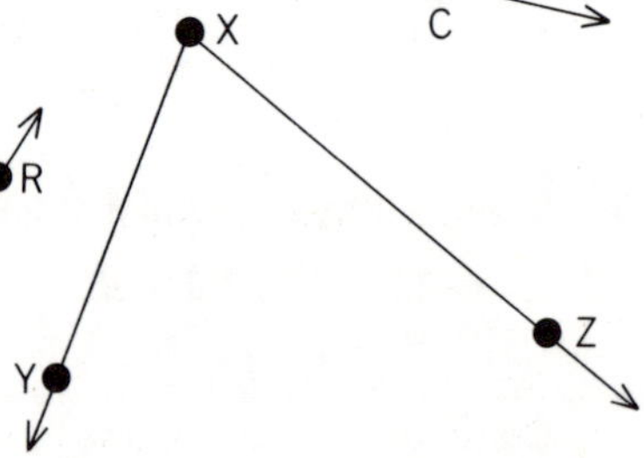

An angle formed by two rays $\overrightarrow{OP}$ and $\overrightarrow{OQ}$, which are in opposite directions on the same line, is equal to 2 right angles. The measure of angle POQ is 2 right angles, or, in shorthand, m($\angle$ POQ) = 2 right angles.

Cut out a circular piece of paper, fold it into halves and then into quarters. Open up your circle and pencil along the folds.

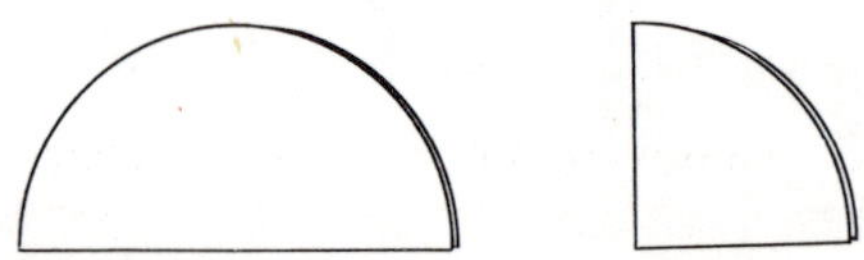

3. How many right angles can you see on your circle?
4. How many right angles make this angle?

- A **full revolution** is made up of four right angles.

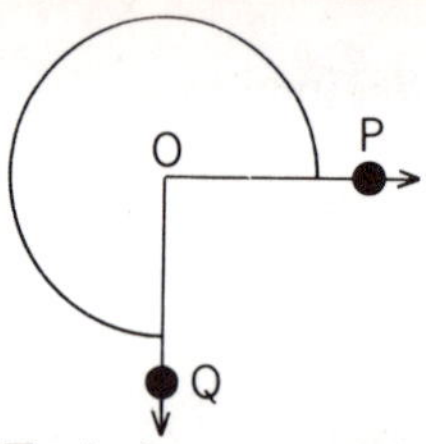

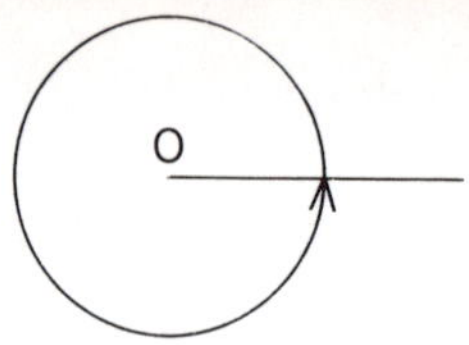

To help you study angles make the following simple aid.
Cut out a semicircle of tracing paper.
Fold it as in diagram **A** and then as in diagram **B**.
Open up your paper and pencil in the creases.
Angle 3 is $1\frac{1}{2}$ right angles.

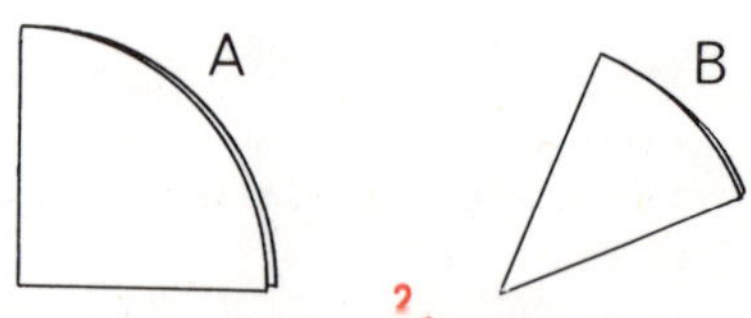

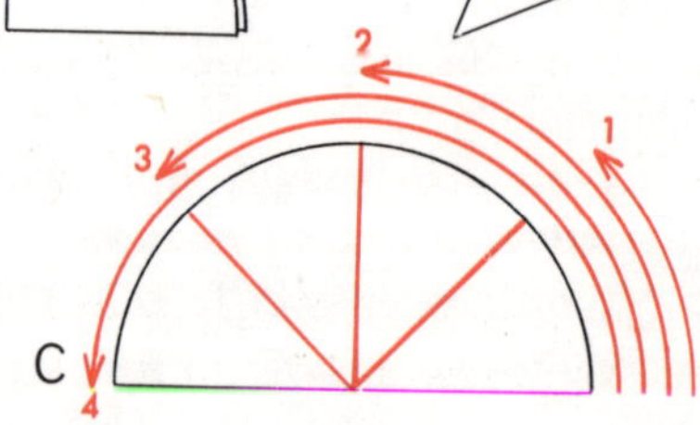

5. How many right angles, or what fraction of a right angle, is each angle marked 1; 2; 4?

Let's Try

1. By fitting your tracing paper over angle PQR check that:
∠ PQR is less than 1 right angle;
∠ PQR is greater than $\frac{1}{2}$ right angle.

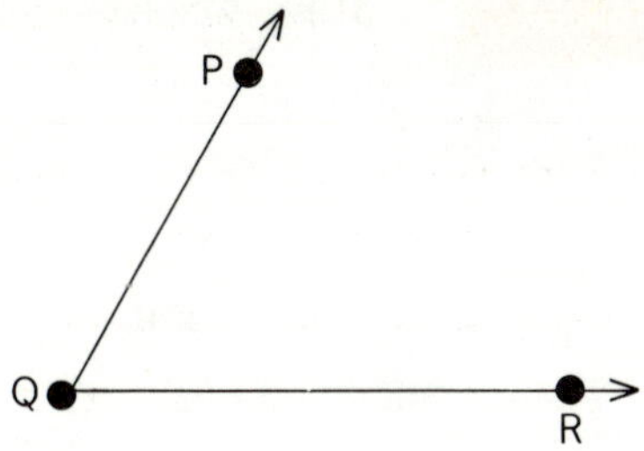

2. Name these angles:

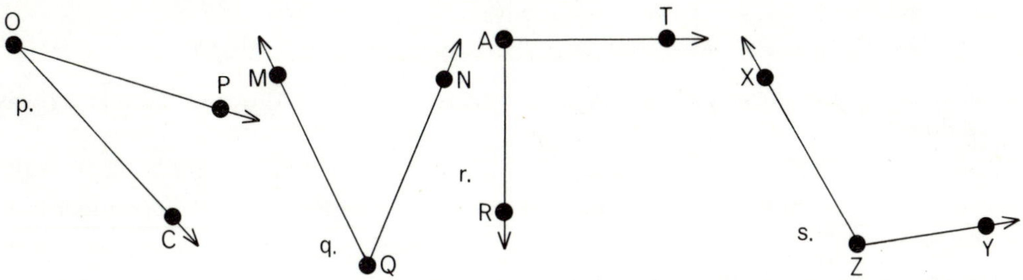

3. Using your tracing paper to help you, write from {p, q, r, s} the subset of angles which are:

   a. right angles
   b. smaller than a right angle
   c. greater than a right angle
   d. greater than $\frac{1}{2}$ a right angle

4. Pair off the angles which are the same size. For example G and H are the same size.

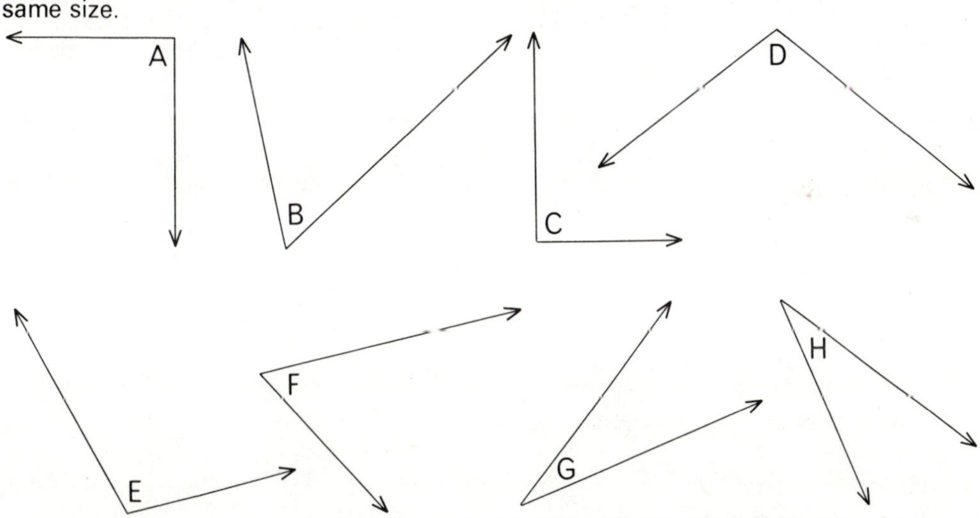

5. Draw and label as ∠ ABC an angle which is greater than 1 right angle.

6. Draw and label as ∠ PQR an angle which is less than $\frac{1}{2}$ right angle.

7. Draw and label as ∠ DEF an angle which is less than 1 right angle but greater than $\frac{1}{2}$ right angle.

8. If you turn left, turn left again and left again, how many right angles have you turned through?

Let's Discover **Juno Arithmetic**

Juno Jim has to go to school. Once he has learned his numbers he can start using operations.

1. Add the following base five numbers for Jim:

a.	b.	c.	d.	e.
2	2	3	114	234
+2	+3	+3	+ 13	+232
___ (5)	___ (5)	___ (5)	___ (5)	___ (5)

Change all Jim's numbers to base ten and see if you are right.

Did you notice how Jim carries to the **'fives'** column and to the **'twenty-fives'** column?

2. Jim has learned how to subtract. Can you do this work for him? Remember that in **base five**

$4+1 = 10_{(five)}$ so $10-1 = 4_{(five)}$ and $10-4 = 1_{(five)}$

a.	b.	c.	d.	e.
4	10	43	124	231
−2	− 3	−14	− 30	− 42
___ (5)	___ (5)	___ (5)	___ (5)	___ (5)

Change the numbers to base ten and check your answers.

3. Jim is coming to our world and he will have to change his numbers to ours. He needs to complete the following sentences and follow through this new method of setting out:

a. Five units make a five.
There are . . . fives in 113.

b. Five fives make a . . .

c. Five twenty-fives make a . . .

5)113
5) 22 (fives) rem. 3 (units)
4 (twenty-fives) rem. 2 (fives)
$113_{(10)} = 423_{(5)}$

4. Change these base ten numbers to base five: 23, 24, 40, 75, 110, 140.

A friend of Jim's came up with a bright idea. He said, "Why not draw a graph so that you can check your answers?"
Here is part of the graph which Tom drew.

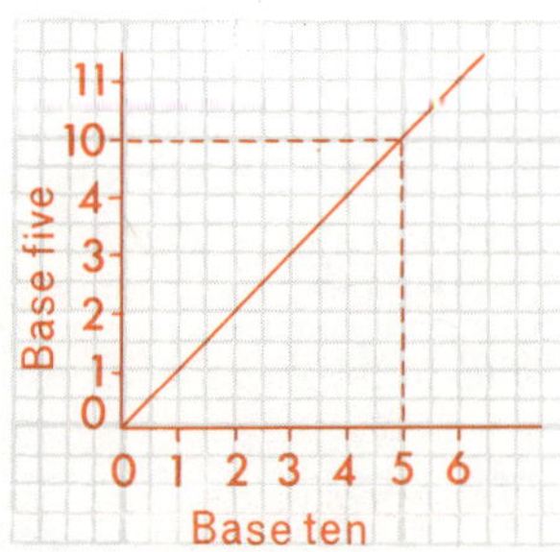

5. Draw your own graph on 2-mm squared graph paper.
Mark the horizontal axis 0 to 50 and the vertical axis 0 to $200_{(five)}$

Let's Try

1. a. Use your graph to change these base ten numbers to base five:
10, 11, 21, 30, 33, 39, 41, 49

   b. Use your graph to change these base five numbers to base ten:
20, 21, 30, 100, 102, 120, 140, 144

2. a. Copy this addition table and fill in the missing numbers.

base five

+	0	1	2	3	4
0					
1					
2				10	
3					12
4					

   b. Copy this subtraction table and fill in the missing numbers.

base five

−	0	1	2	3	4
4		3			
10					
11					
12				4	
13					4

3. Complete these addition sums in base five:

a. 22 + 14 = ___(5)

b. 131 + 22 = ___(5)

c. 231 + 134 = ___(5)

d. 432 + 32 = ___(5)

e. 244 + 201 = ___(5)

4. Copy these subtraction sums and complete them in base five:

a. 40 − 2 = ___(5)

b. 32 − 13 = ___(5)

c. 113 − 22 = ___(5)

d. 222 − 133 = ___(5)

e. 321 − 114 = ___(5)

5. Use the method of dividing by five to change these base ten numbers to base five:

a. 22 b. 43 c. 79 d. 121 e. 120 f. 200 g. 262

6. Here are two magic squares containing base five numbers. The sum of the numbers in any column or any row is $30_{(5)}$.
Fill in all the missing numbers in base five using the numbers:
1, 2, 3, 4, 10, 11, 12, 13, 14.

4	14	
3	10	

12		3
2		4

Let's Discover **Time**

Here is a picture of a very old clock.
The clock has a pendulum which swings from side to side.

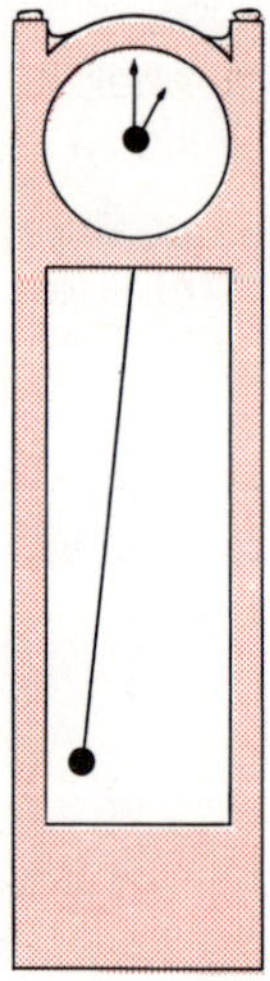

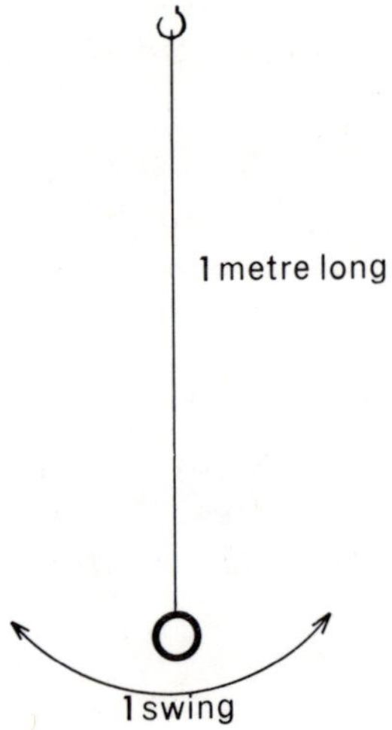

1. **a.** Find a small but heavy weight such as a metal ring. Attach a piece of strong thin thread to the weight. Hang the weight from a nail or hook so that the length of the thread is **1 metre**.

b. Work with a partner. Time 20 swings in order to check that 1 swing takes 1 second.

You now have a pendulum which we call a **seconds pendulum** because each swing takes 1 second.

2. Complete these sentences:

a. 60 swings of my pendulum takes . . . seconds or . . . minutes.

b. 60 × 60 swings of my pendulum takes . . . seconds or . . . minutes or 1 hour.

Here is a picture of a modern clock which shows 10.05

10.05 is 5 min past 10

Now the clock shows 04.54

04.54 is 54 min past 4
or 6 min to 5

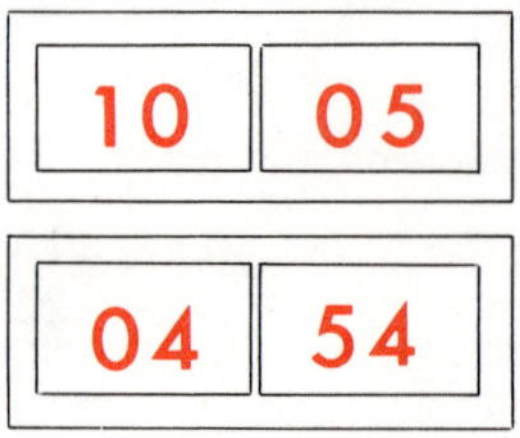

Here is a series of pictures taken every $\frac{1}{4}$ hour from midday, 12.00, onwards

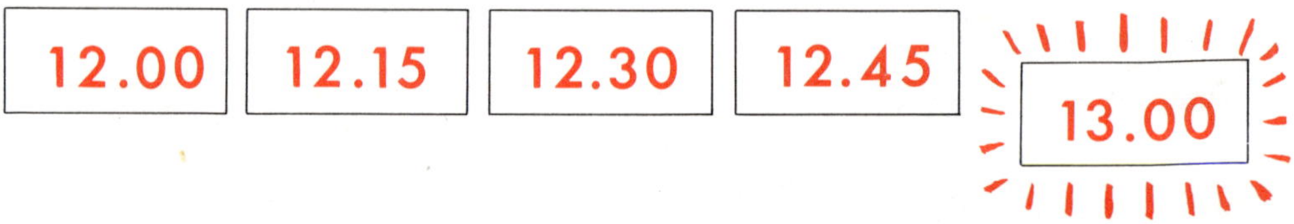

The last picture shows 13.00 because it is a 24-hour clock.
13.00 is the same time as 1 p.m.

This chart shows 12-hour clock and 24-hour clock times.

	a.m.					p.m.				
12-hour clock	8	9	10	11	12	1	2	3	4	5
24-hour clock	08.00	09.00	10.00	11.00	12.00	13.00	14.00	15.00	16.00	17.00

3. Write the 12-hour clock and the 24-hour clock time for:

a. breakfast b. lunch c. tea
d. your morning break e. your end of school day

- 24-hour clock times are used for airline time-tables and train time-tables so that a.m. and p.m. times are not confused.

You wouldn't like to arrive for a plane at 1 p.m. only to discover that it went at 1 a.m.

Here is a train time-table.

Train A takes

h	min
10	05
7	55

2 h 10 min for the journey

train	A	B	C
Crewe (depart)	07.55	09.20	12.05
London (arrive)	10.05	11.30	13.55

4. Which train leaves in the afternoon and arrives at 1.55 p.m. or 5 min to 2?

Let's Try

1. Here is a piece of a bus time-table.
How long does it take to travel from:
a. Belton to Ceton?
b. Axton to Ceton?
c. Axton to Deton?
d. Copy and complete the missing part of the time-table for the 10.15 and 13.10 buses.
e. Write the 13.10 time-table using 12-hour clock times.

Depart	Time		
AXTON	9.30	10.15	13.10
BELTON	10.10	10.55	
CETON	10.55		
DETON	11.40		

2. a.

h	min
4	53
+2	37

b.

h	min
3	27
−1	43

c.

h	min
2	30
×	5

d. $1\frac{3}{4}$ h ÷ 7
= . . . min

3. Say "Jack Robinson" 100 times and divide the number of seconds taken (use your pendulum) by 100.
How long does it take you to say "Jack Robinson"?
Repeat this with other sayings, such as "Quick as a flash".

Let's Discover **Parallel Lines**

Here are two lines, $\overleftrightarrow{AB}$ and $\overleftrightarrow{CD}$.
It is quite easy to see that if enough of each line is drawn, they intersect at E.

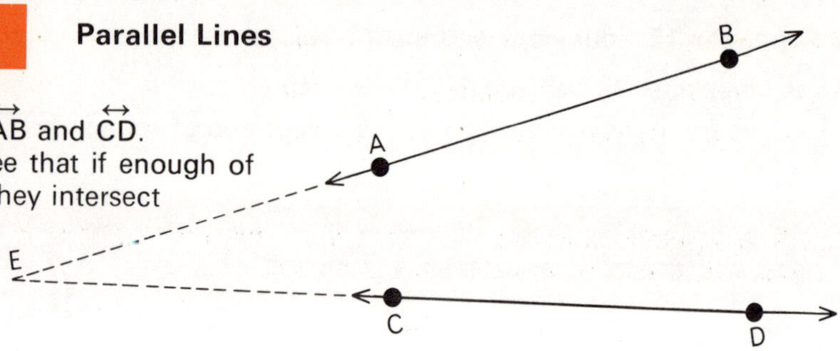

1. Do you think that these two lines $\overleftrightarrow{PQ}$ and $\overleftrightarrow{RS}$ would ever intersect?

   No matter how much of the lines you draw they will never intersect.

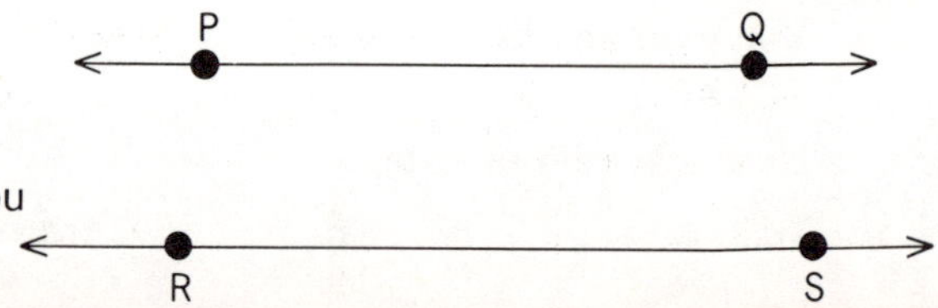

- Two lines which never intersect are called **parallel lines**.

Here are two pairs of parallel line segments.
$\overline{EF}$ is parallel to $\overline{GH}$, and
$\overline{JK}$ is parallel to $\overline{LM}$.

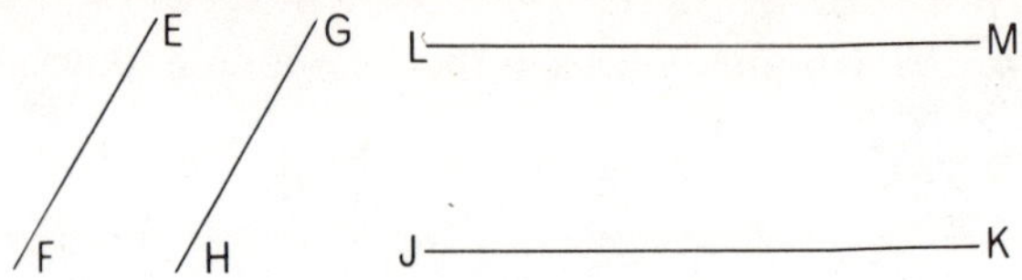

2. Look for parallel lines or line segments in these pictures.

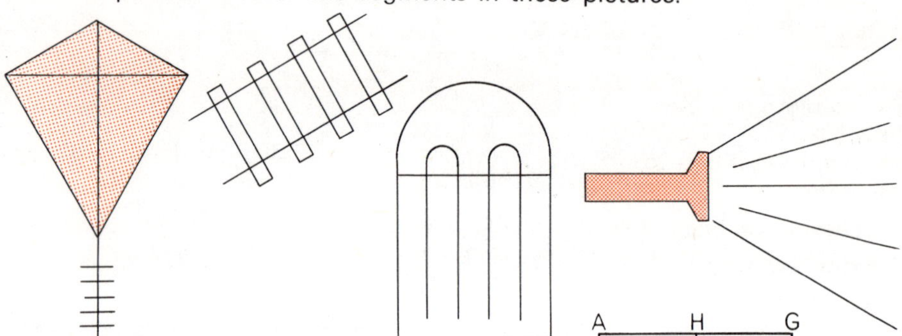

3. This is a picture of a window. There are many pairs of parallel line segments on this window.

   For example, $\overline{AB}$ is parallel to $\overline{HO}$,

   and $\overline{OF}$ is parallel to $\overline{CD}$

   Write down some more pairs of parallel line segments.

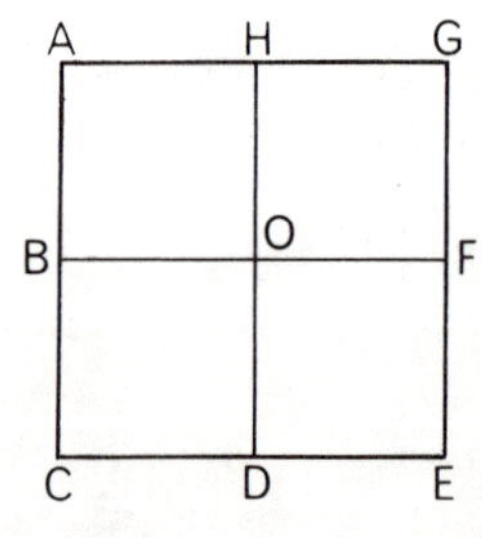

Here is a quadrilateral which is made up of two pairs of parallel sides.

$\overline{AB}$ is parallel to $\overline{DC}$

$\overline{BC}$ is parallel to $\overline{AD}$

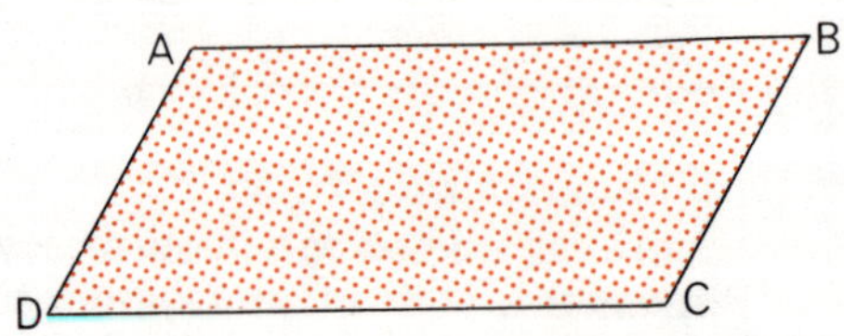

4. Use a ruler to measure the sides of ABCD.

Is $m(\overline{AB}) = m(\overline{DC})$, and $m(\overline{BC}) = m(\overline{AD})$?

The shape ABCD is called a **parallelogram**.

- A quadrilateral with opposite sides parallel and equal in length is a **parallelogram**.

Let's Try

1. From this set name each pair of parallel line segments.

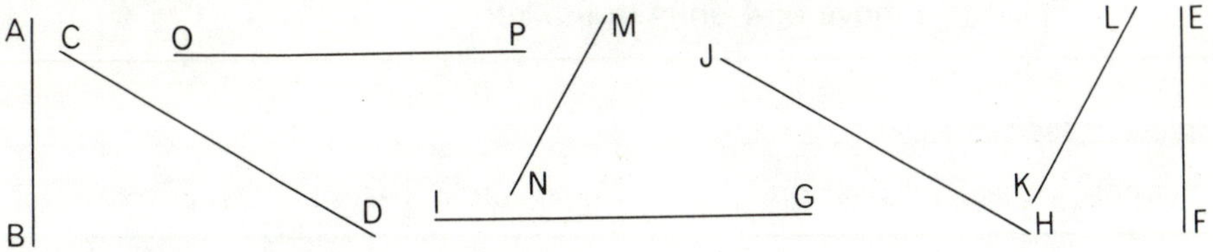

2. Which of these shapes are parallelograms?

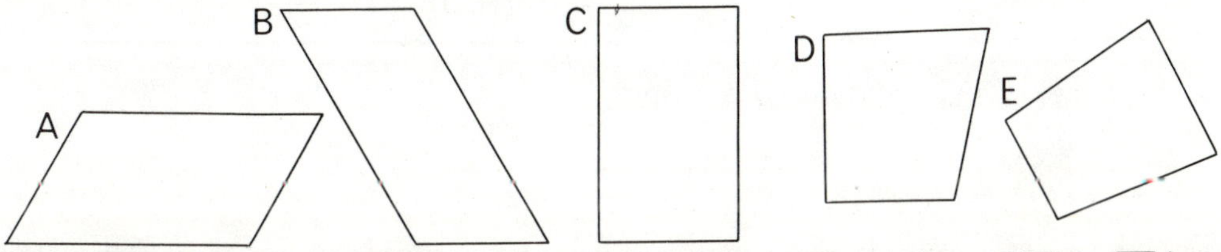

3. Cut out four strips of card like these and pin them with
A joined to A, B joined to B,
C joined to C, and D joined to D.
What shape have you made?

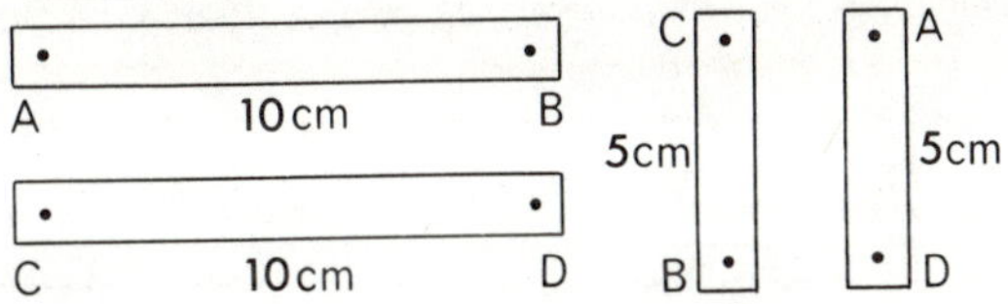

4. Draw and cut out of card any parallelogram you like. Build up a tessellation with your parallelogram.

5. Draw and cut out a parallelogram like this one.
Tear off the four angles after marking them A, B, C, and D.

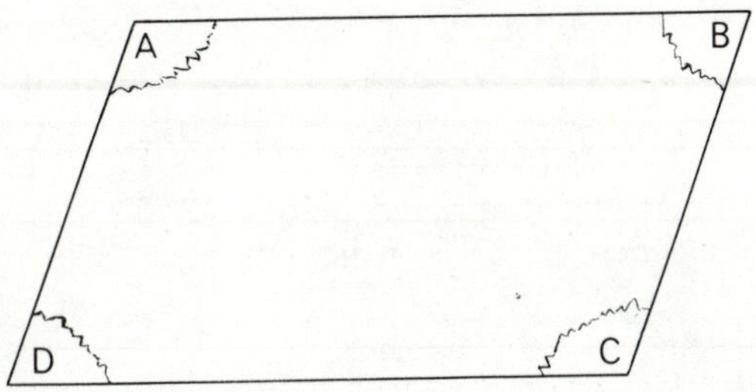

a. What can you say about angles A and C?

b. What can you say about angles B and D?

c. Fit angles A and B together.
How many right angles do A and B add up to?

d. Fit angles A, B, C, and D together.
How many right angles do they add up to?

Let's Discover

Another Use for Ordered Pairs

Here is part of a post office sorting box.

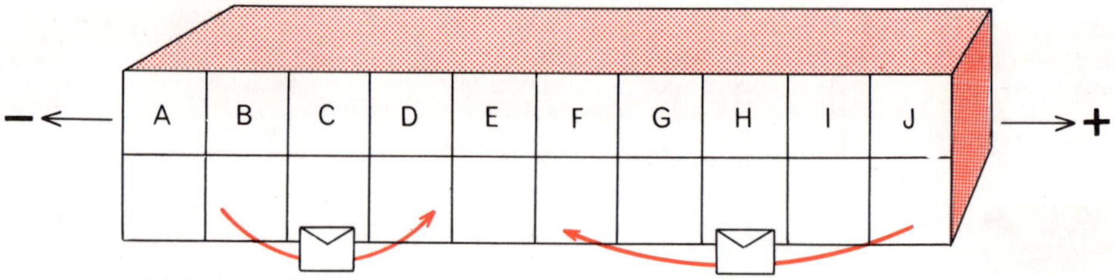

The instruction (D, B) says move a letter **from B to D**.
(D, B) is a move of 2 units to the right.
The instruction (F, J) says move a letter **from J to F**.
(F, J) is a move of 4 units to the left.

1. Complete these tables.

Instruction	Units	Direction
(E, C)		Right +
(F, D)		
(J, H)	2	

Instruction	Units	Direction
(C, E)		Left −
(D, F)	2	
(H, J)		

In example **1** you should have found:

- The units travelled can be the same for **different** instructions.
- The units travelled **and** direction can be the same for **different** instructions.

2. Does (C, E) produce the same result as (E, C)?

In a pair, such as (C, E), **order** matters, so we call this an **ordered pair**.

Let's change our boxes from	A	B	C	D	E	F	G	H	I	J
to	1	2	3	4	5	6	7	8	9	10

The instruction (5, 2) says move from 2 to 5 which is 3 units.
Since (5, 2) is a movement of 3 units towards + write (5, 2) = $^{+}3$.

3. Use the new boxes to help you complete:

a. (6, 2) = $^{+}$. . . **b.** (5, . . .) = $^{+}2$ **c.** (4, 1) = . . .

On the numbered boxes, (2, 5) says move from 5 to 2.
Since (2, 5) is a movement of 3 units towards − write (2, 5) = $^{-}3$.

4. Use the new boxes to help you complete:

a. (2, 6) = $^{-}$. . . **b.** (3, . . .) = $^{-}2$ **c.** (1, 4) = . . .

Let's Try

1. Here is a motorway with access points 1 to 15. The access points are 1 unit of length apart.

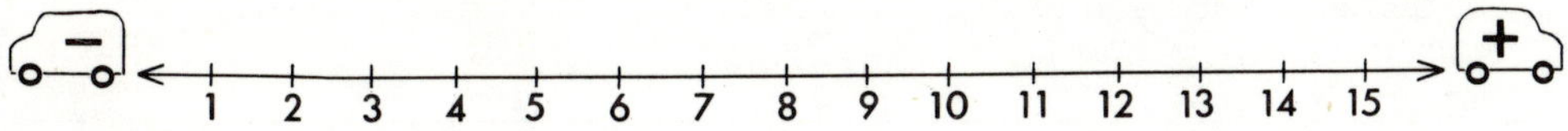

Copy and complete this table. **a** has been done for you.

	Move to	From	Units travelled	+	−
a.	8	6	2	√	
b.	6	4			
c.	4	6			
d.	11	15			
e.	10		2		√
f.	9		4	√	

2. Use the number boxes from the opposite page to help you complete the following:

a. (2, 4) = $^{-}$. . . **b.** (4, 2) = . . . **c.** (2, 7) = . . .
d. (7, 2) = . . . **e.** (9, 2) = . . . **f.** (3, 3) = . . .
g. Copy this table and find the replacement for △ in each ordered pair.

	(4, △)	(△, 2)	(△, 6)	(3, △)	(6, △)
Units and sign	$^{+}2$	$^{+}2$	$^{-}3$	$^{-}4$	2

3. Which element is the odd one out in each of the following sets?

a. {(1, 4), (2, 5), (6, 3), (4, 7)} **b.** {(4, 1), (5, 2), (3, 6), (7, 4)}
c. {(1, 5), (2, 6), (6, 2), (4, 8)} **d.** {(6, 1), (8, 3), (10, 5), (9, 3)}

4. Complete these ordered pairs:

a. (4, . . .) = $^{-}2$ **b.** (2, . . .) = (4, 9)

5. | (0, 2) = $^{-}2$ | Write four other pairs which equal $^{-}2$.

Let's Discover: Symmetry

1. Fold a piece of paper into halves like this.
   Keep the paper folded and carefully cut out any shape you like.
   Start at the crease and finish somewhere else on the crease.

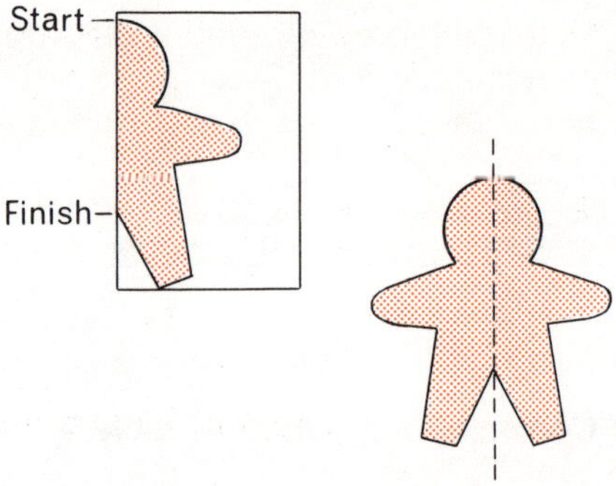

2. Open up your shape like this.
   The crease separates the new shape into two matching parts.

Your shapes might look like these:

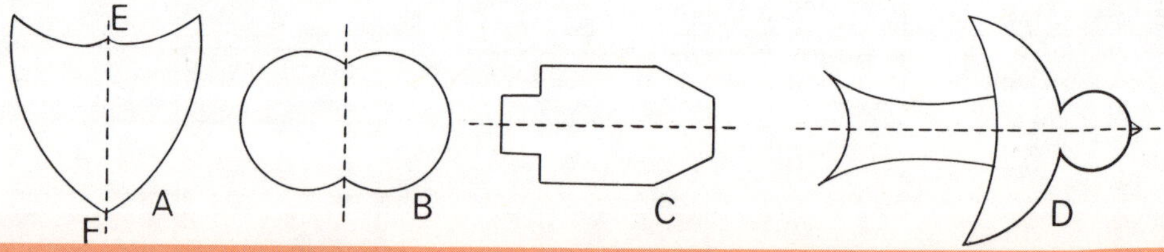

- Each shape is **symmetrical** about a line segment.

$\overline{EF}$ divides shape A into two parts which match.
Shape A is symmetrical about line segment $\overline{EF}$.

$\overleftrightarrow{EF}$ is called a **line of symmetry** of shape A.

3. Run your finger along the line of symmetry for each shape: B, C, and D.

4. How many lines of symmetry has this shape?
   $\overline{PQ}$ is one line of symmetry.
   $\overline{RS}$ is another line of symmetry.

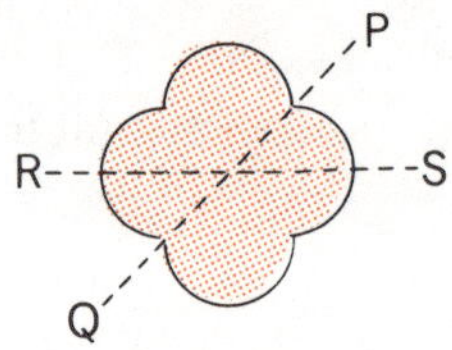

5. Has this quadrilateral any lines of symmetry?

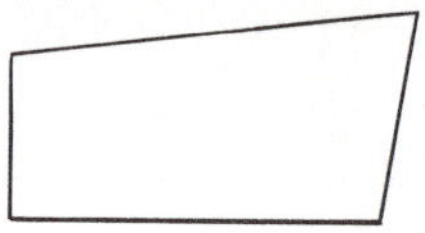

6. Two lines of symmetry are shown on this square. Draw two more lines of symmetry.

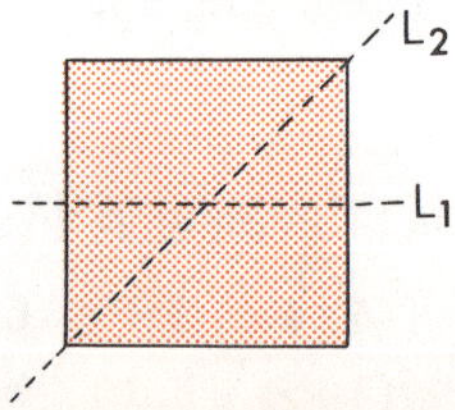

- Some quadrilaterals are symmetrical and some are not.

A rectangle has two lines of symmetry: L_1 and L_2.

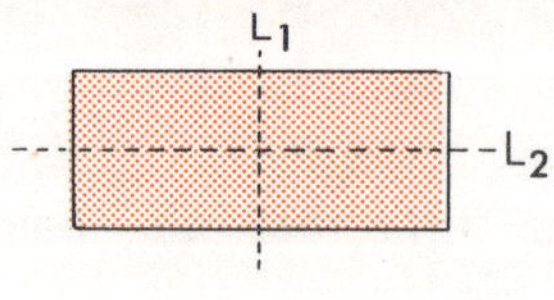

Let's Try

1. How many lines of symmetry has each of these shapes?

a. 

b. 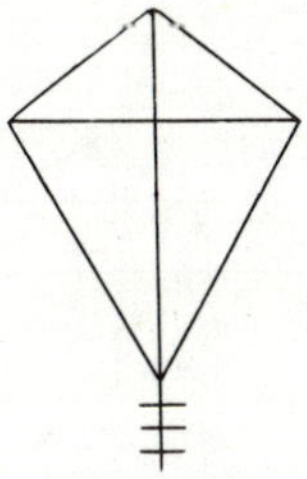

c. 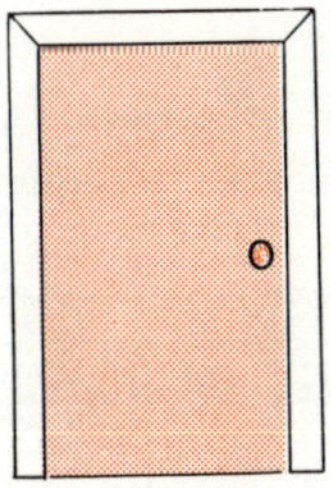

d. 

e. 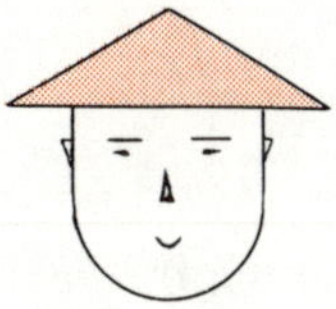

f. 

g. 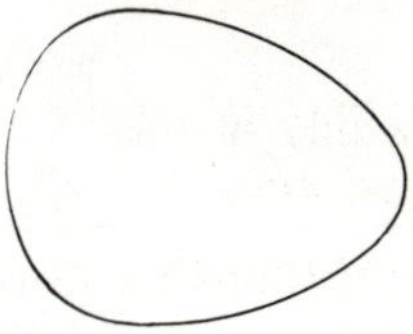

h. 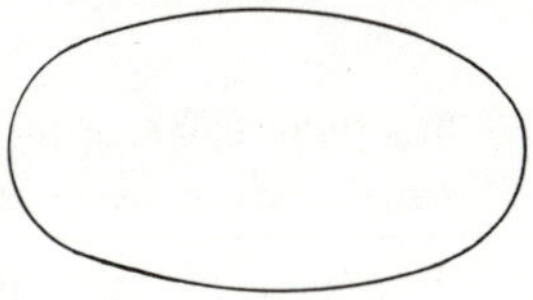

2. Draw the two lines of symmetry on a rectangle like this one.

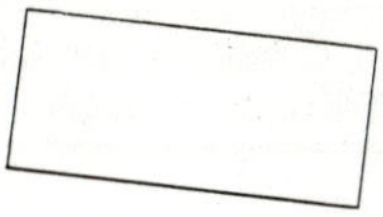

3. Has this parallelogram any lines of symmetry?

4. How many lines of symmetry has this quadrilateral?

5. How many lines of symmetry has this quadrilateral?

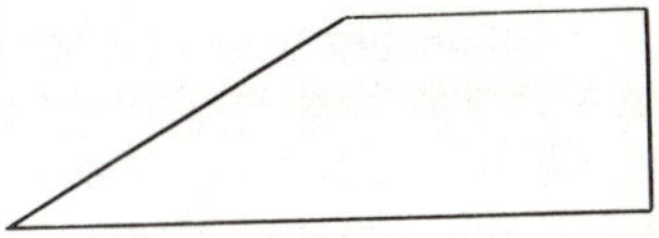

6. Which of these quadrilaterals have:
   a. no lines of symmetry?
   b. only two lines of symmetry?

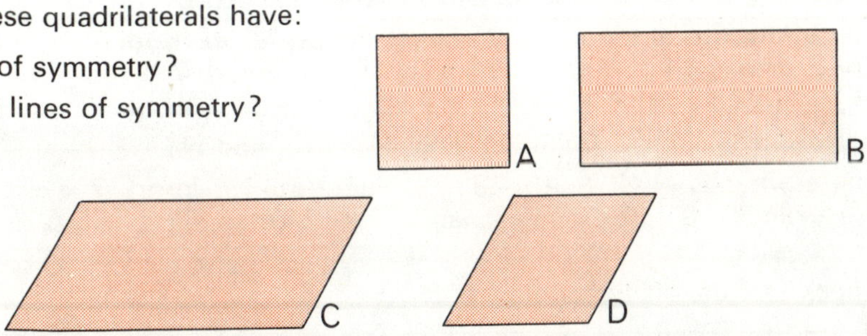

7. The following shapes are incomplete but are symmetrical about the dotted line. Copy and complete the shapes.

a. 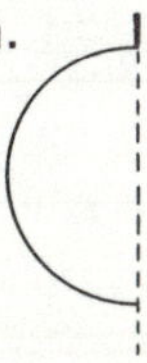

b. 

c. 

d. 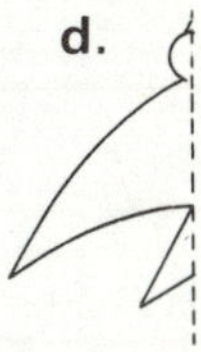

Let's Discover **Modular Arithmetic**

Here is a machine which automatically divides a number by 7.

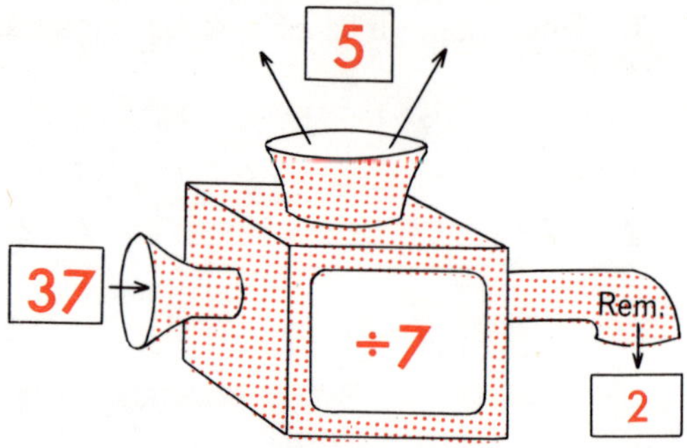

1. 37 has been fed into the machine which tells us that there are . . . sevens in 37 and . . . remainder.

2. Feed each element of this set into the machine and write the remainder in each case. {9, 51, 65, 72, 100}

3. If today is Monday what day of the week will it be in 45 days time?

If you found the answer to question **3** quickly you probably thought:

45÷7 = 6 rem. 3   In 6 weeks or 42 days time it will be Monday. Another 3 days will make it Thursday.

4. The elements of this set all have a remainder 3 when divided by 7. {3, 10, 17, 24, . . .} Write the next three elements.

5. Write the first five numbers which have a remainder of:

a. 1   **b.** 2   **c.** 3   **d.** 4   **e.** 5   **f.** 6   **g.** 0
when divided by 7.

You are beginning to learn about a new kind of mathematics in which remainders are very important. We call this mathematics, **modular arithmetic**.

The **elements** of each of the following sets are **equal in modulo 7 (mod. 7)** since they have the same remainder when divided by 7.

The **elements** of {0, 7, 14, 21, . . .} are **equal in mod. 7**
For example, 14 = 0 (mod. 7), and 49 = 0 (mod. 7)

The **elements** of {1, 8, 15, 22, . . .} are **equal in mod. 7**
For example, 8 = 1 (mod. 7), and 50 = 1 (mod. 7)

The **elements** of {2, 9, 16, 23, . . .} are **equal in mod. 7**
For example, 9 = 2 (mod. 7), and 51 = 2 (mod. 7)

6. Check that the elements of:

a. {3, 10, 17, 24, . . .}   **b.** {4, 11, 18, 25, . . .}
c. {5, 12, 19, . . .}   **d.** {6, 13, 20, . . .}
are equal in mod. 7.

7. We can add and multiply in mod. 7.
For example, $4+4 = 8 = 1 \pmod{7}$ and $4\times4 = 16 = 2 \pmod{7}$.
Check and complete these addition and multiplication tables in mod. 7.

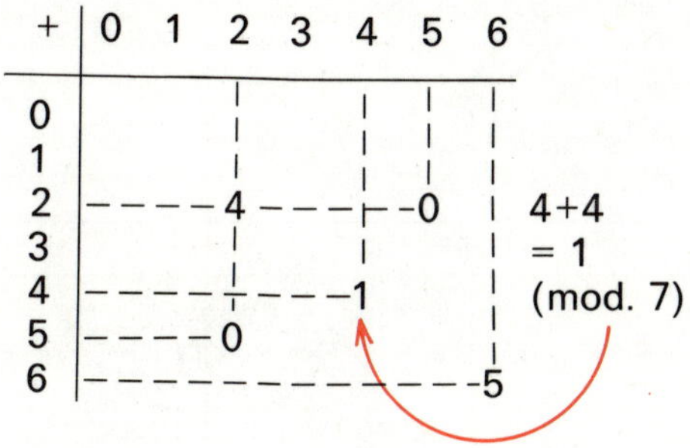

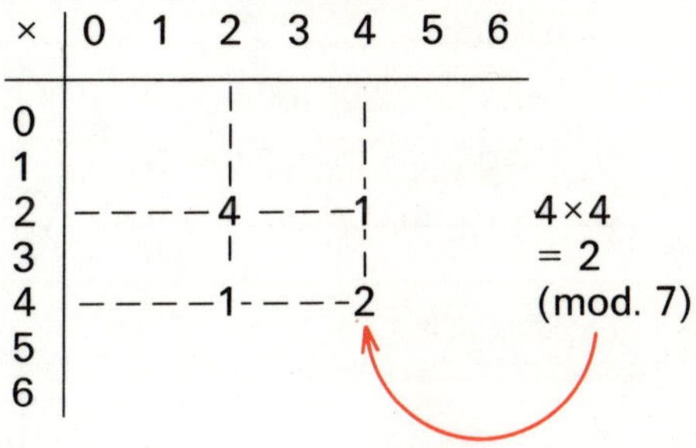

Let's Try

1. Find the element of {0, 1, 2, 3, 4, 5, 6} which is equal to:

a. 49 **b.** 73 **c.** 89 **d.** 140 **e.** 139 **in mod. 7.**

2. Find the smallest replacement for △ in:

a. 4+5 = △ (mod. 7) **b.** 4×5 = △ (mod. 7) **c.** 6+6 = △ (mod. 7)
d. 6×6 = △ (mod. 7) **e.** 8+9 = △ (mod. 7) **f.** 8×9 = △ (mod. 7)
g. 7×7 = △ (mod. 7) **h.** 7+7 = △ (mod. 7) **i.** 54×7 = △ (mod. 7)

3. Complete the following sentences:

a. 63 days from **today** it will be . . . day.
b. If I turn a 7-channel TV switch forwards 5 clicks I will be on channel . . . if I started on channel 4.
c. In 13 months time the month will be . . .
d. On my 12-hour clock it is 10 o'clock. In 7 hours time it will be . . . o'clock.

Let's Discover **More Symmetry**

1. Find as many plane shapes as you can which have:

a. one line of symmetry;

b. two lines of symmetry;

c. three lines of symmetry.

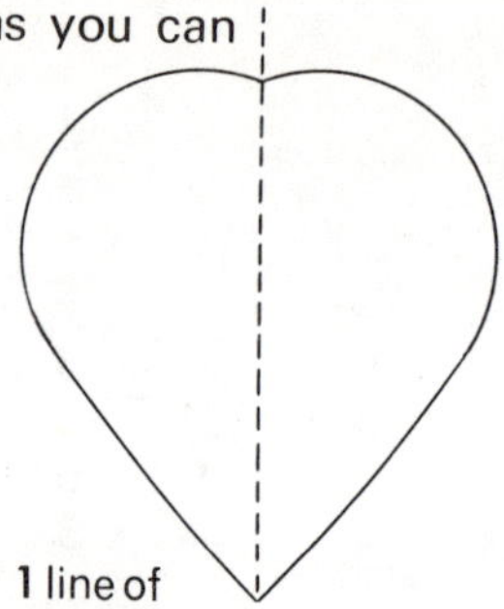

1 line of symmetry

7 lines of symmetry

An equilateral triangle has **three lines of symmetry**.

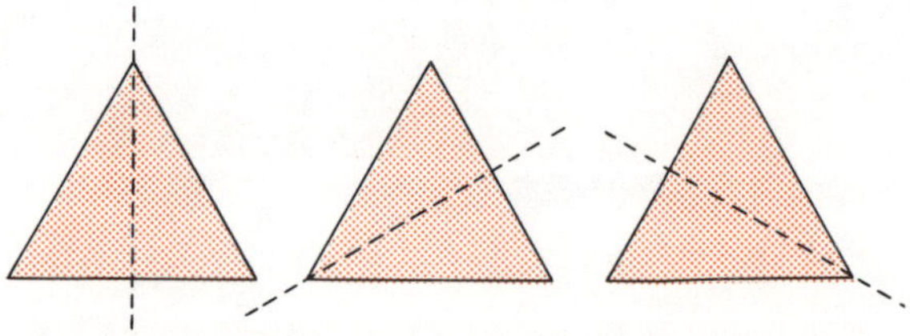

2. a. Cut out an equilateral triangle and fold it along its axes of symmetry.

b. Do your lines of symmetry meet at a point?

- The point at which two or more lines of symmetry meet is called the **centre of symmetry**.

Catherine-wheels are usually circular.

3. a. What would the pattern from a square catherine-wheel look like when revolving?

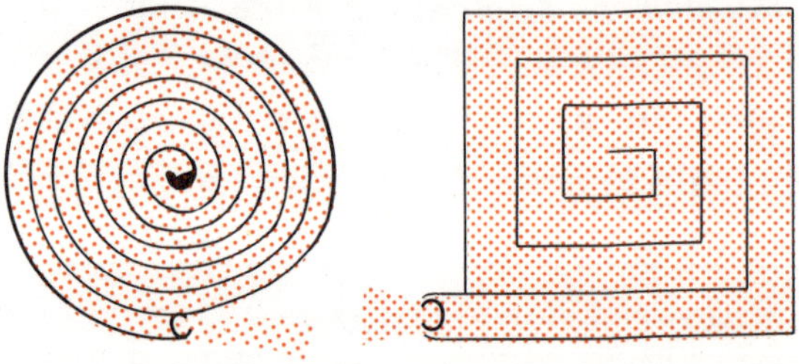

b. Pin a square piece of cardboard through its centre of symmetry. Spin the square round quickly.

c. Pin a square through its centre of symmetry onto a clean piece of paper.
Hold the square still and draw round it.
Rotate your square slightly and draw round it again. Repeat this until you have completed a circular pattern.

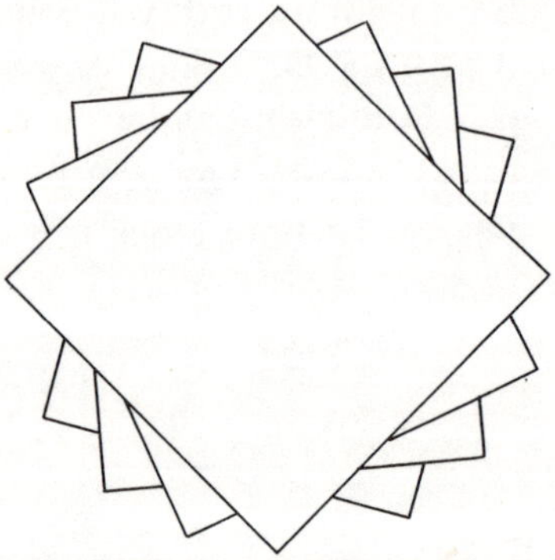

Patterns similar to this can be drawn using shapes which have a centre of symmetry.

Let's Try

1. Shape A is a square. Cut out a square piece of paper and fold it so that lines like the ones on this square appear as creases.

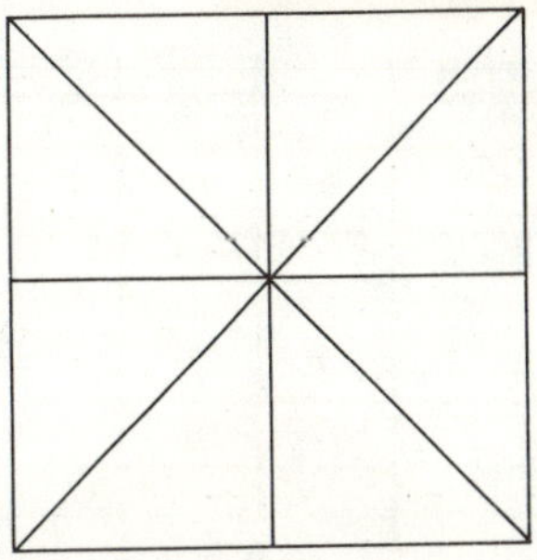

A

2. Shape B is a rectangle.
Repeat question **1** with a rectangle.

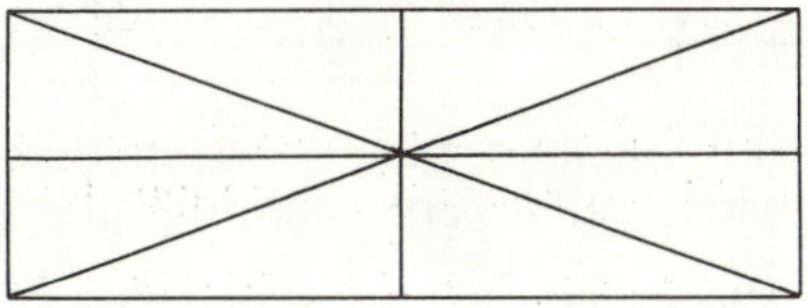

B

3. a. Does a square have a centre of symmetry?
   b. Does a rectangle have a centre of symmetry?

4. a. How many lines of symmetry has each of these snowflakes?
   b. Has each snowflake a centre of symmetry?

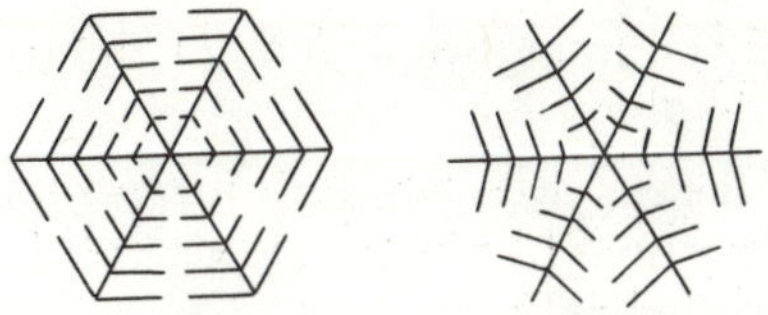

5. Look carefully at these shapes.
   a. How many lines of symmetry has each shape?
   b. Which of these shapes has a centre of symmetry?
   c. Cut out a shape like C. Pin it to a board through its centre. Rotate it and draw a circular pattern.

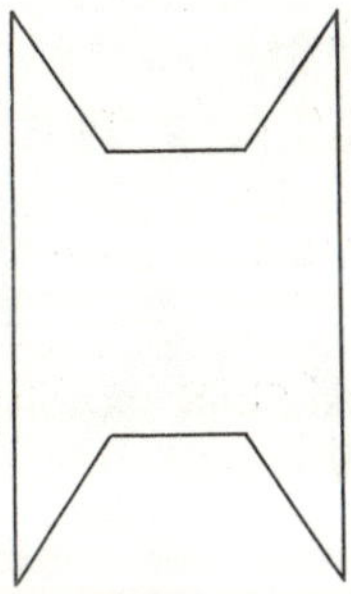

C

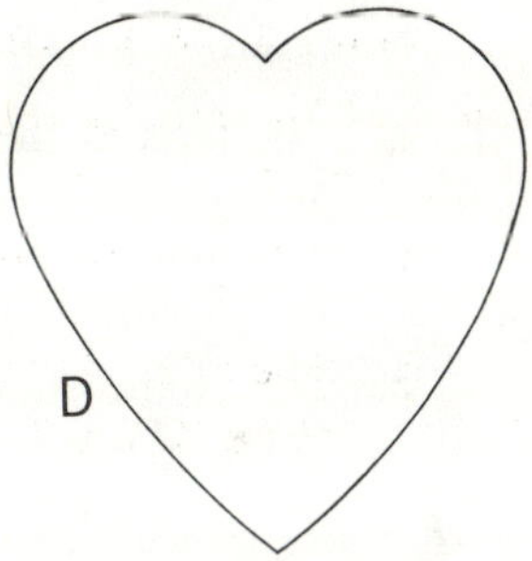

D

6. Which of these flags have:
   a. one line of symmetry?
   b. two lines of symmetry?
   c. a centre of symmetry?

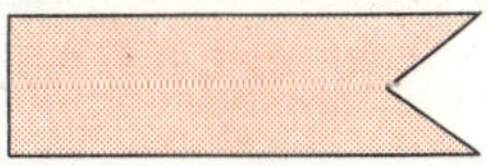

E

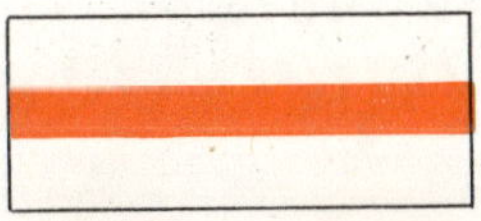

F

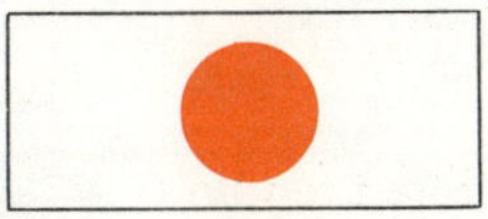

G

H

7. How many lines of symmetry can be drawn on:
   a. shape J?
   b. shape K?
   c. shape L?

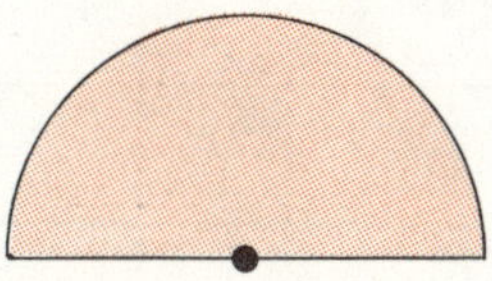

J

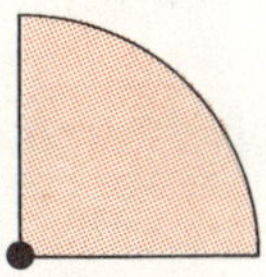

K

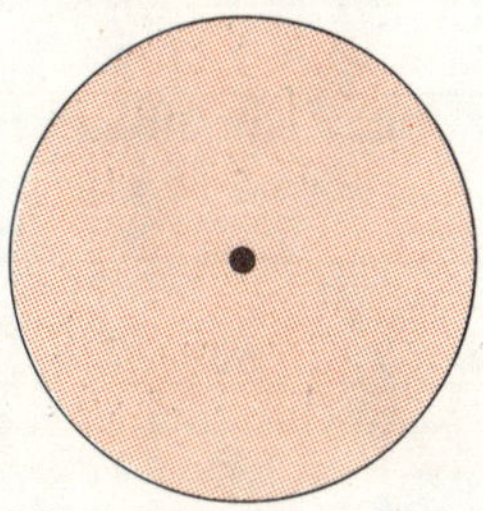

L

8. Which one of the shapes, J, K, and L, has a centre of symmetry?

Let's Discover

More Area

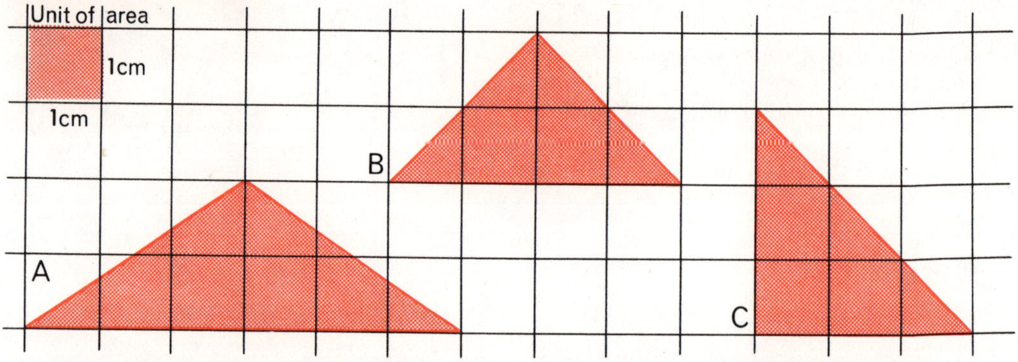

1. Cut out shapes like these shaded triangles from squared paper. Make one cut on each triangle and rearrange the two pieces into a square or a rectangle.

The area of regions which are not rectangular can be measured. Triangle A has an area of 6 cm^2.

2. By counting squares and parts of squares find the area of the other two triangular regions.

This diagram shows a piece of cardboard from which a letter L is going to be cut.
The standard unit, one small square, is the **square centimetre**.

3. **a.** The area of the card is "x" cm^2.
x = . . .

b. The area of the card to be cut away is "y" cm^2. y = . . .

c. The area of the L-shaped region is "x−y" cm^2. x−y = . . .

- You have found the area of the red region by **subtracting** the area of the white region from the area of the piece of card.

Here is a letter T which is cut from card.

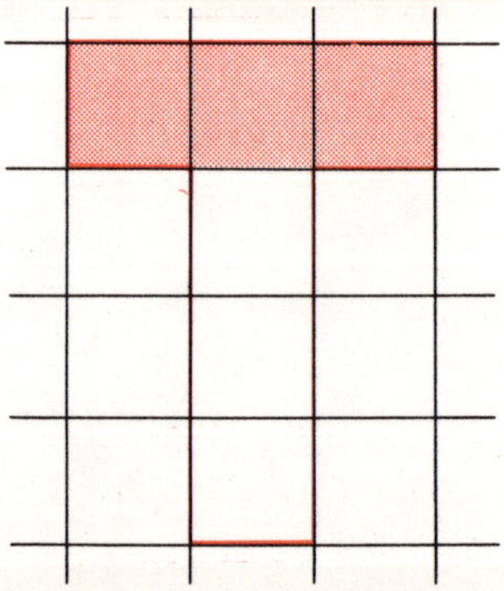

4. **a.** The measurement of the red region is x cm^2. x = . . .

b. The measurement of the white region is y cm^2. y =

c. The area of the T-shaped region is x+y cm^2. x+y = . . .

- You have found the area of the T-shaped region by **adding** the area of the red region to the area of the white region.

Let's Try

1. Here are some pictures of walls which have been tiled. Complete the statements.

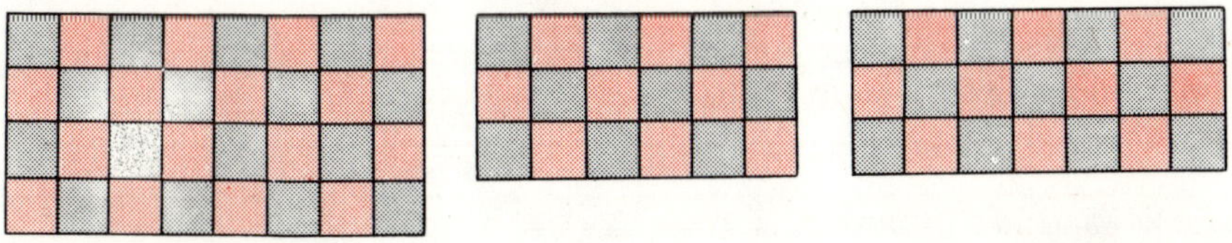

a. Number of tiles = . . .
Number of red tiles = . . .
Number of grey tiles = . . .

b. Number of tiles = . . .
Number of red tiles = . . .
Number of grey tiles = . . .

c. Number of tiles = . . .
Number of red tiles = . . .
Number of grey tiles = . . .

2. Draw these regions on to squared paper. The numbers on the edges of each region show lengths in cm.
By adding areas find the area of each region.

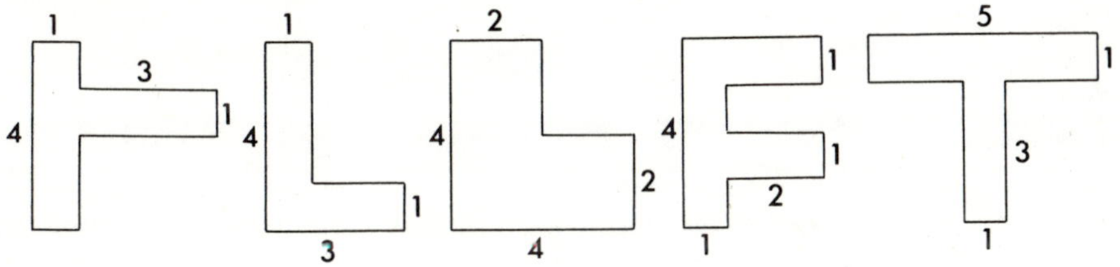

3. Draw these regions on to squared paper. The numbers on the edge of each region show lengths in cm.
By subtracting areas find the area of each region.

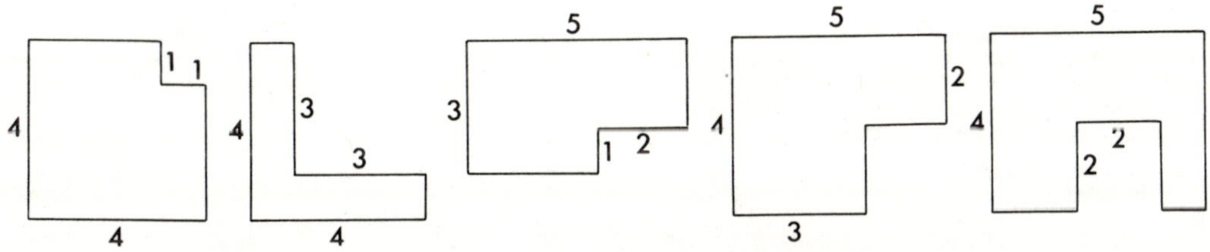

4. Here is a picture of a carpet.

a. What is the area of the carpet?

b. Find the area of the red region.

c. Work out the total area of the grey regions.

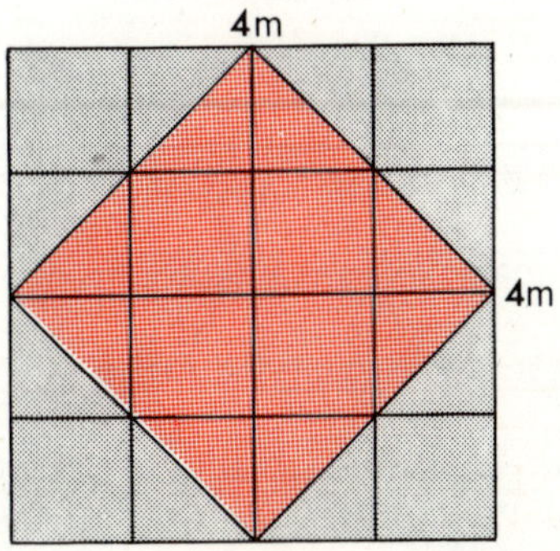

Let's Discover **Volume**

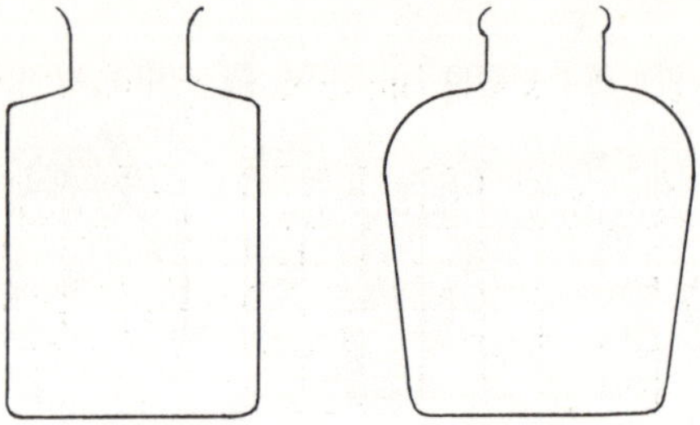

1. Find two different containers which you think are about the same size.
2. Fill one container with sand (or water) and then pour your sand into the other container without spilling any.
3. Have you any sand left in your first container?
4. Do you need some more sand to completely fill your second container?
5. Which of your containers holds the most sand?

Your larger container has more space in it.
A measure of the space is called the **volume**.

Here are some things which occupy space. They are called solids.

You might think that some of these things are not very solid but in mathematics any object which **occupies space** is called a **solid**. A measure of the space occupied by each solid is called its volume.

6. Make a collection of cups and jars which seem to have about the same volume. Fill each container with sand and pour it into a tall bottle. Make a mark where the sand comes to. Put your containers in order of size.

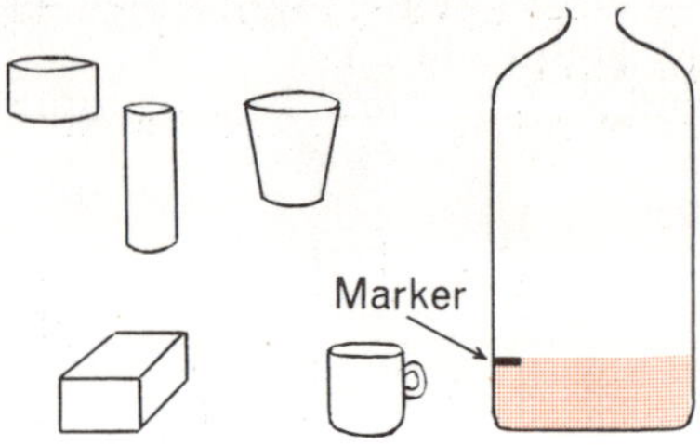

Sugar is sometimes bought in lumps.
Look at a lump of sugar carefully.

7. Is the lump a **mathematical** solid?
8. Is each face a square?
9. Is each edge a straight line segment?

- A solid like this is called a **cube**.

10. Make a list of all the objects you can find which are cubes.

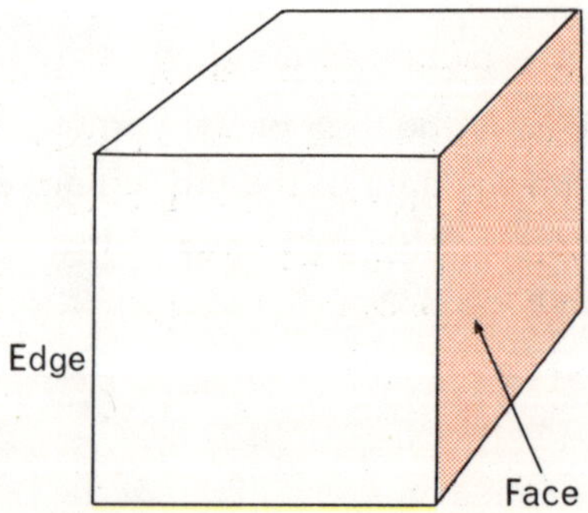

Let's Try

1. Sixteen cubes can be arranged into blocks like these.

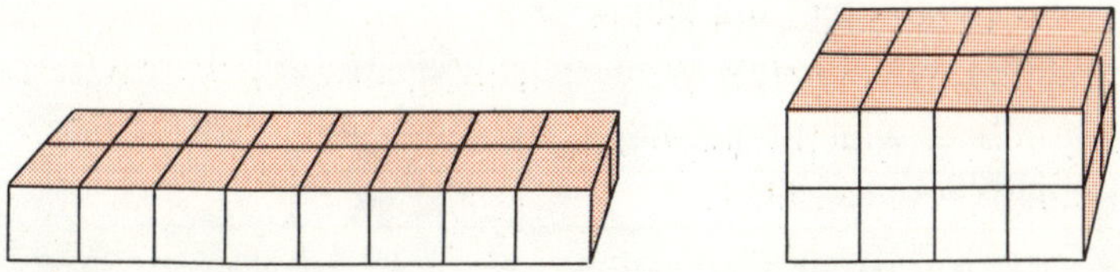

a. Arrange sixteen cubes in as many other blocks as you can.
b. What can you say about the volume of each block?

2. How many of these cubes have the same volume as the following solids?

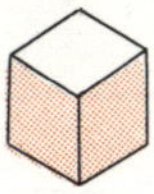

a.

b.

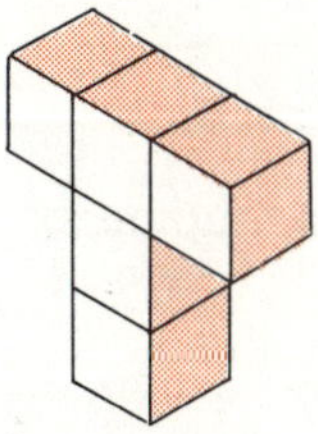

c.

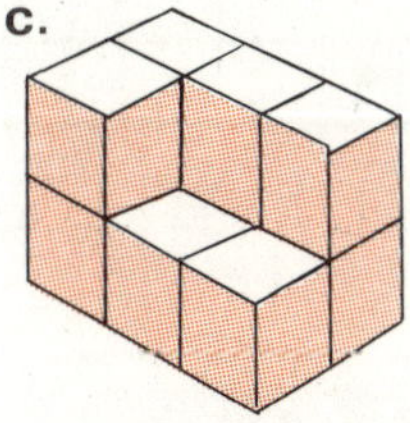

d.

e.

f.

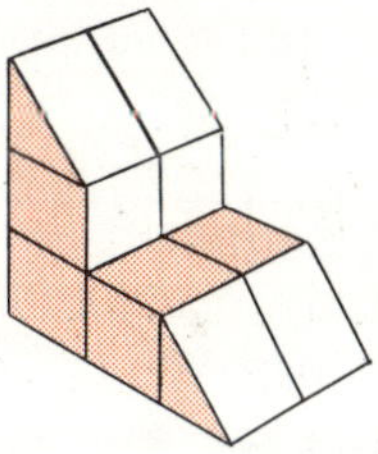

3. Make three plasticine cubes of the same size. Cut two cubes as in the diagram. Write the replacement for each △.

	Volume		Volume
a.	A+B	=	△×A
b.	B+A	=	△×B
c.	A+A	=	△×C
d.	A+B	=	△×C

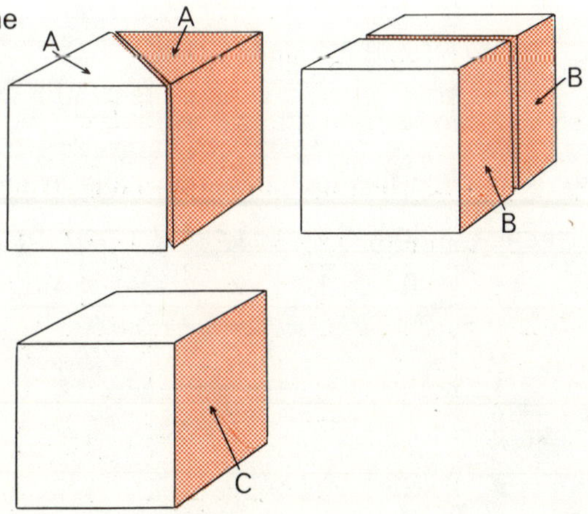

Let's Discover

More about Remainders

1. Here is a series of numbers: 1, 4, 7, 10, 13, 16, .

a. Write the next four numbers in the series.

b. Change each number in the series to its remainder in mod. 7.

2. Here is a series of numbers which have been changed to their remainders in mod. 7.

Series	0	2	4	6	8	10	12	14
mod. 7	0	2	4	6	1	3	5	0

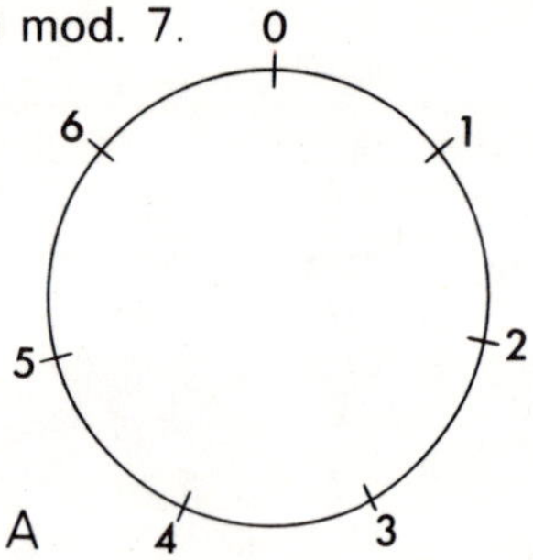

A

Copy circle A and join up the mod. numbers starting 0 → 2 → 4 . . . as in circle B.

3. Complete this table:

Series	1	4	7	10	13	16	19	22
mod. 7	1	4	0	—	—	2	—	—

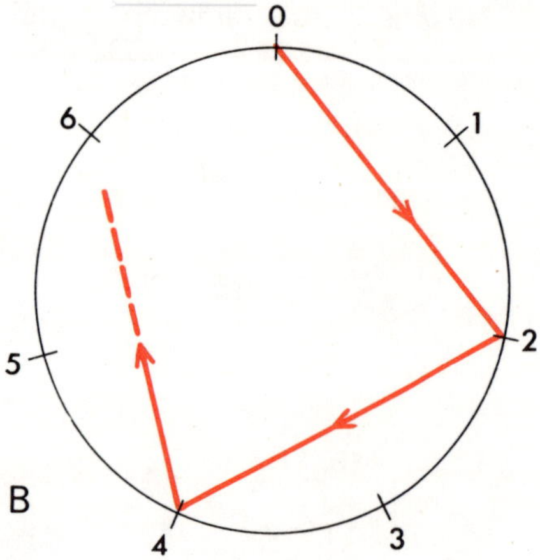

B

Copy circle A and join up your mod. numbers starting 1 → 4 → 0 → . . .

You have learned how to graph series of numbers in mod. 7.

So far you have learned about mod. 7 arithmetic.
In mod. 7 your divisor was 7 and remainders were {0, 1, 2, 3, 4, 5, 6}.

There are many other modular systems. For example, **in mod. 5** your set of remainders will be {0, 1, 2, 3, 4}.

In mod. 5

25 = 0 (mod. 5) since 5)25
                                        5 rem. 0

36 = 1 (mod. 5) since 5)36
                                        7 rem. 1

4. Complete this multiplication table in mod. 5.

For example, 4×3 = 12 = 2 (mod. 5)

×	0	1	2	3	4
0					
1					4
2					3
3					
4				2	

Let's Try

1. Find the element of {0, 1, 2, 3, 4} which is equal to:

   a. 49 **b.** 73 **c.** 89 **d.** 95 **e.** 94 **in mod. 5.**

2. Complete these mod. 5 tables. Put a × on each 0 and a red × on each 1. Shade each 2 in grey and each 3 in red.

Mod. 5

+	0	1	2	3	4
0	0	1			
1		2	3	4	
2					
3					
4		0			3

Mod. 5

x	0	1	2	3	4
0	0				
1		1		3	4
2		2			
3		3			
4					

3. Repeat question **2** for mod. 7. Set up your own squares. Choose another form of shading for all the 5s and all the 6s. Look at the patterns formed by your shading.

4. Here are some series. Complete each table and graph the series on a circle like the one underneath it.

a.

Series	1	5	9	13	17	21	25	29
mod. 7								

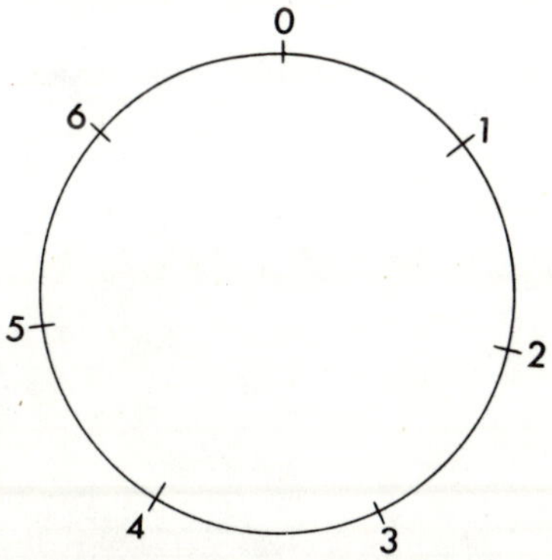

b.

Series	0	2	4	6	8	10
mod. 5						

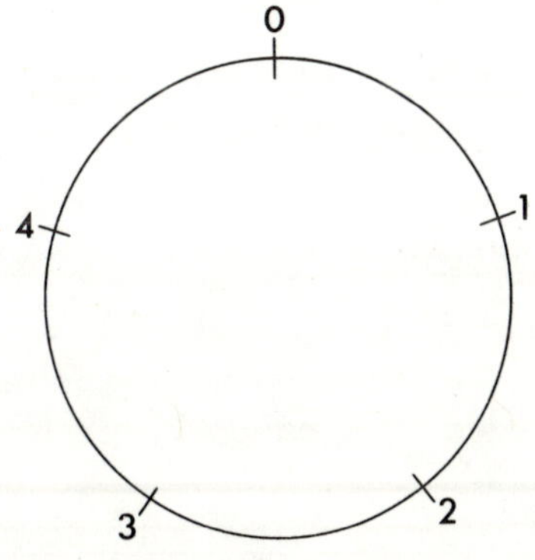

5. Complete these tables and draw a graph for each series.

a.

Series	0	3	6	9	12	15	18	21
mod. 7								

b.

Series	0	3	6	9	12	15
mod. 5						

6. What do you notice about the graphs of **4a** and **5a**? Is the same true about **4b** and **5b**?

Let's Discover **Cuboids**

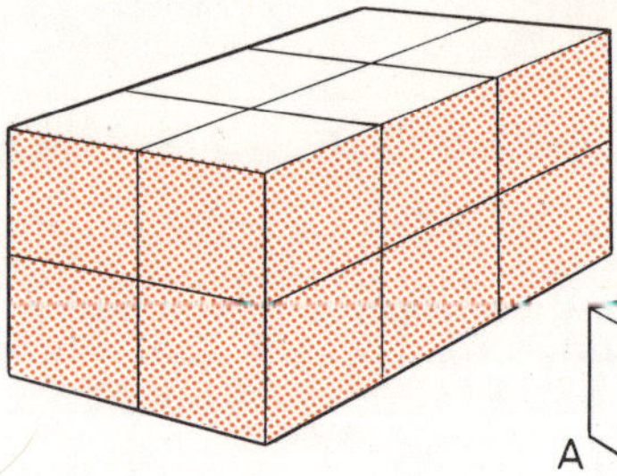

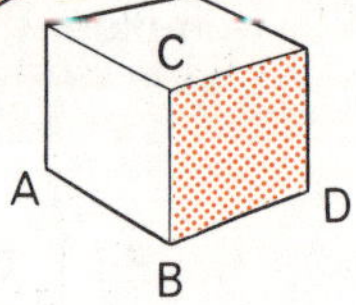

This block is made up of 1-cm cubes.

1. a. How many 1-cm cubes have been used?
   b. How many 1-cm cubes must be removed in order to make the block into a 2-cm cube?

A cube has m($\overline{AB}$) = m($\overline{BC}$) = m($\overline{BD}$)

2. Find a brick which is a cube and run your finger along the 12 edges which are equal in length.

   The measures on these blocks are shown as cm.

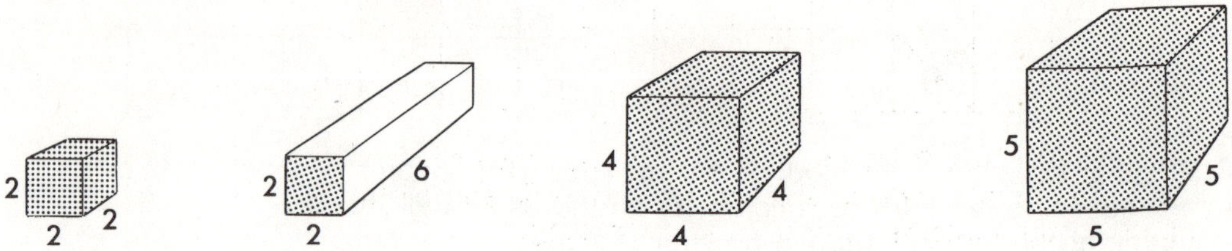

3. a. What is the shape of each shaded face?
   b. Which of these blocks is not a cube?

   If one of the measures: m($\overline{AB}$), m($\overline{BC}$) or m($\overline{BD}$) is not the same as the other two then the block is called a **cuboid**.

   If the three measures: m($\overline{AB}$), m($\overline{BC}$), and m($\overline{BD}$) are all different then the block is a **cuboid**.

   Here is a cuboid which is

   6 units long, 4 units wide, and 3 units deep.

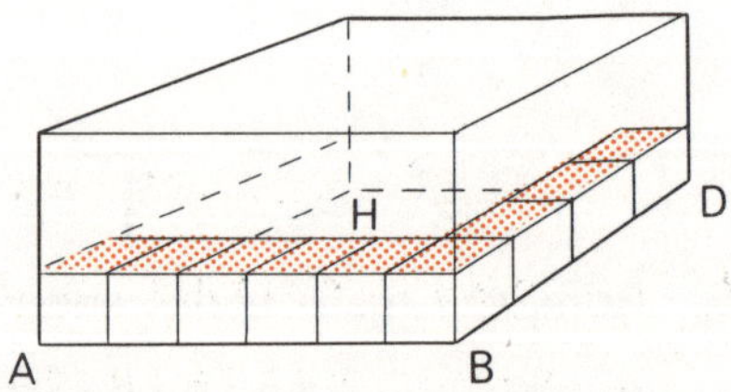

4. l units of volume have been placed end to end along edge AB. l = . . .

5. b units of volume have been placed end to end along edge BD. b = . . .

6. How many units of volume can be placed on the lower base rectangle ABDH?

   Since the cuboid is 3 units deep, three layers of cubic units are needed to build the cuboid.

Each layer of the cuboid will be made up of 6×4 units.

The number of **cubic units** needed to build the **cuboid** is $6 \times 4 \times 3 = 72$.

- If the unit of length is **1 cm**, the unit of volume is a **1-cm cube**. Volume is then measured in **cubic centimetres (cm^3)**.

The volume of a cuboid 6 cm long, 4 cm wide, and 3 cm deep is 72 cm^3.

Let's Try

1. This box is 5 cm long, 4 cm wide, and 3 cm deep. Suppose we start filling the box with 1-cm cubes.

   a. How many cubes will form the bottom layer?

   b. How many cubes will form the second layer?

   c. How many layers will there be?

   d. How many cubes will be needed to fill the box?

   e. What is the volume of the box in cm^3?

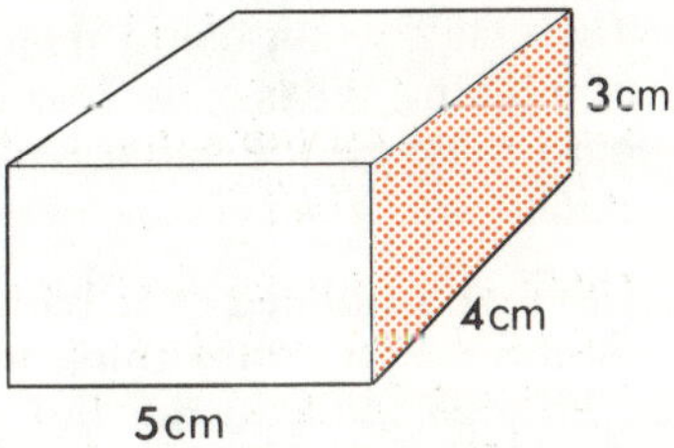

2. Better Building bricks cost 5p per packet of 10 bricks. How many bricks are needed to build a wall:

   a. 10 bricks long, 4 high, and 1 thick?

   b. 10 bricks long, 10 high, and 2 thick?

   c. 20 bricks long, 4 high, and 2 thick?

   d. How many packets are needed for each of the three walls?

   e. What is the cost of the bricks for each wall?

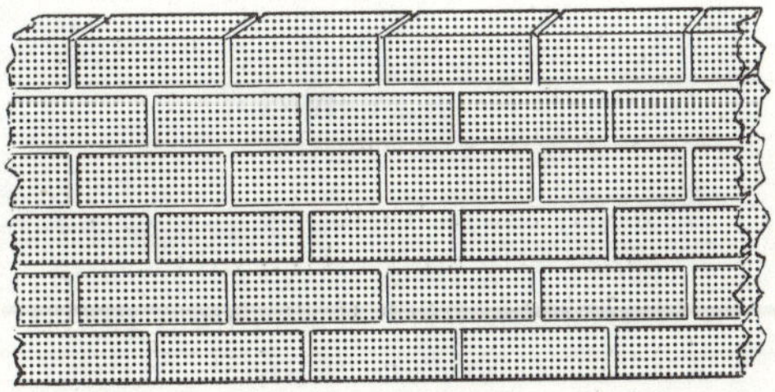

3. How many 1-cm cubes will fit into a box which is:

   a. 10 cm long, 10 cm wide, and 10 cm deep?

   b. 5 cm long, 5 cm wide, and 5 cm deep?

   c. What is the volume of each of the two boxes?

Let's Discover

How to Use Weights

The teacher asked Jane to cut a 10-cm square from a piece of card which measured 14 cm by 14 cm.
The diagram shows how Jane cut the card.

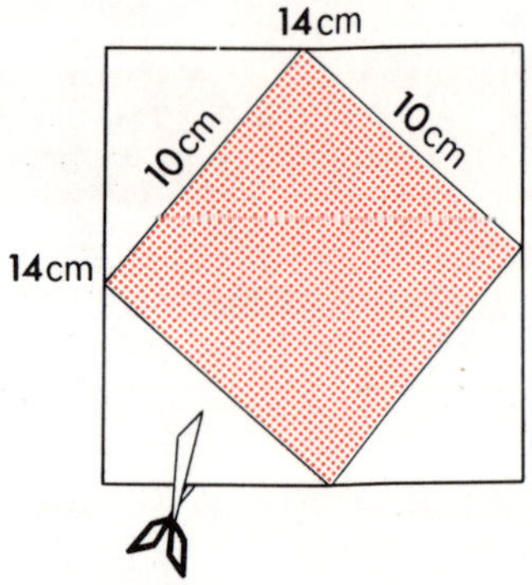

1. Complete the following:

   a. The area of a 10-cm square
   $= 10 \times 10 \text{ cm}^2$
   $= \ldots \text{cm}^2$

   b. The area of a 14-cm square = . . .

2. The teacher told Jane that nearly half the piece of card had been wasted. How many cm^2 had been cut away?

   Instead of working out the area of the pieces of card we could **compare their weights**.
   The square which Jane cut out has a larger area and so is heavier than the wasted card.
   Take care when using this method of comparing areas. The two pieces of card must be the same type and thickness. Why?

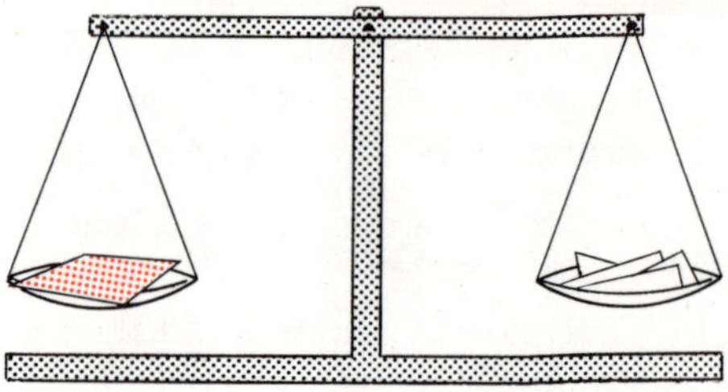

   Have you noticed that banks weigh money? You might think that this is a strange thing to do, but it saves the clerk a great deal of time.

3. **A 1p coin weighs 3·6 g**

   Complete this working:

   10 pennies weigh . . . g
   100 pennies weigh . . . g

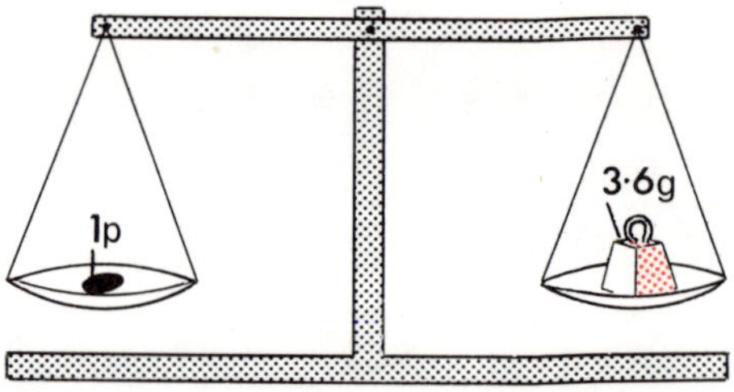

   The bank clerk knows that a bag contains £1 in pennies if its **net weight** is 360 g.

   A shopkeeper brought in a bag of copper and said, "This bag contains **£1 in pennies**." It weighed **360 g**.
   Later the bank clerk opened the bag and found that it contained a 2p coin and 98 pennies.

4. How much money was there in the bag? Did the bag weigh correctly for £1?

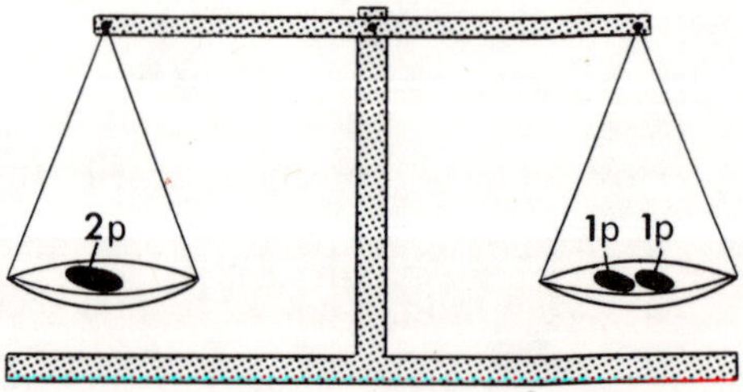

- It is useful to remember that a 2p coin weighs twice as much as a 1p coin.

5. Complete this working:
A 2p coin weighs $3 \cdot 6 \times 2$ g = . . . g

Let's Try

1. How many 2p coins amount to £1?
How many 2p coins weigh the same as one hundred 1p coins?

2. A 2p coin weighs 7·2 g. Find the weight of the following numbers of 2p coins.

a. 10 **b.** 5 **c.** 20 **d.** 7

3. Complete the following:

		Weight in g
a.	Ten 1p coins	
	Five 2p coins	
	Total: . . . p	. . . g

		Weight in g
b.	Forty 1p coins	
	Twenty 2p coins	
	Total: . . . p	. . . g

c. $3 \cdot 6\,g \times 20$ = . . . **d.** $3 \cdot 6\,g \times 80$ = . . .

4. A bag contains only 1p and 2p coins. How much money is there in the bag if its net weight is:

a. 360 g **b.** 36 g **c.** 180 g **d.** 3·6 kg

5. **a.** Draw shape A and the boat on a piece of squared card. Copy the boat carefully.
Cut out square A and the boat.

b. Weigh the boat and write its weight in g.

c. Weigh shape A and write its weight in g.

d. Check that the boat balances shape A and complete this statement:
The boat and shape A w . . . the same and have the same a . . .

e. How many of shape A should weigh the same as the piece of card shown in the diagram?

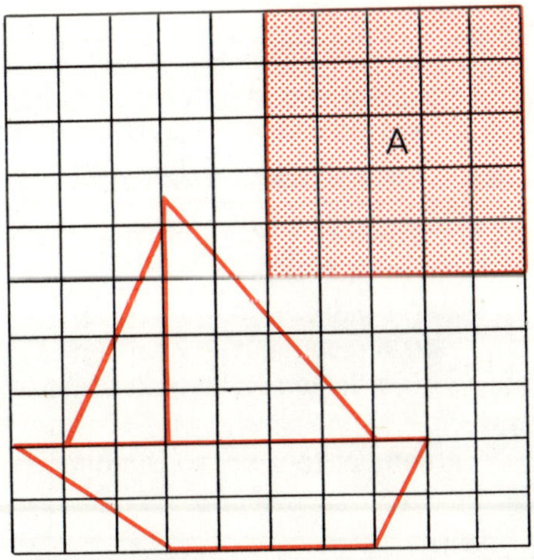

6. Here are some puzzles for you.

a. I have three copper coins in my pocket. Their total weight is 14·4 g. What are the coins?

b. A bag contains twenty 1p coins and weighs 82 g. What is the weight of the bag?

Let's Discover **The Cost**

We have learned how to multiply a decimal by 10 or 100.

1. Complete this example which will help you refresh your memory.

   a. £1·24×10 = . . .   b. $12\frac{1}{2}$p×10 = . . . p = £ . . .

   c. £1·12×100 = . . .   d. £1·$12\frac{1}{2}$×100 = . . .

   There are 21 children coming to Ann's party and she wants to buy each of them a small present costing $5\frac{1}{2}$p.

2. a. What would be the cost of 10 presents at $5\frac{1}{2}$p each?

   b. What would be the cost of 20 presents at $5\frac{1}{2}$p each?

   If we **add** the cost of 1 present to the cost of 20 we will find the cost of 21 presents.

   10 presents at $5\frac{1}{2}$p cost 55p
   10 presents at $5\frac{1}{2}$p cost 55p
   1 present at $5\frac{1}{2}$p costs $5\frac{1}{2}$p
   so 21 presents at $5\frac{1}{2}$p cost $115\frac{1}{2}$p or £1·$15\frac{1}{2}$

   To find the cost of a number of articles, **round off** to the nearest 10.
   In question **2**, 21 was rounded off to 20. The cost of 1 article was **added** to the cost of 20 in order to find the cost of 21 articles.

3. Complete this example in which 32 is rounded off to 30.
   To find the cost of 32 articles at $3\frac{1}{2}$p each:

   10 articles at $3\frac{1}{2}$p cost 35p
   20 articles at $3\frac{1}{2}$p cost . . .p
   2 articles at $3\frac{1}{2}$p cost . . .p
   so 32 articles at $3\frac{1}{2}$p cost . . .p

   Instead of rounding off to the nearest 10 below it is sometimes more convenient to work to the nearest 10 above.
   To find the cost of 18 articles, we find the cost of 20 and **subtract** the cost of 2.

4. Complete this example in which 49 is rounded off to 50 to find the cost of 49 articles at $4\frac{1}{2}$p.

   10 articles at $4\frac{1}{2}$p cost 45p
   40 articles at $4\frac{1}{2}$p cost 180p
   so 50 articles at $4\frac{1}{2}$p cost 225p

   49 articles at $4\frac{1}{2}$p will cost
   225p−$4\frac{1}{2}$p = $220\frac{1}{2}$p
   = £2·$20\frac{1}{2}$

   Remember to work to the nearest ten to make your working as simple as possible.

Let's Try

1. Complete the following statements:

   a. 2p× . . . = £1 b. . . .p×2 = £1 c. . . .p×5 = £1
   d. . . .p×8 = £1 e. $\frac{1}{10}$ of £1 = . . .p f. $\frac{1}{5}$ of £1 = . . .p
   g. $\frac{3}{4}$ of £1 = . . . h. $\frac{3}{4}$ of £2 = . . .

2. Multiply each of the following by 10. (Write the answers in £.)

   a. 15p b. 25p c. 95p d. 125p e. £1·20 f. £0·07$\frac{1}{2}$
   g. 12$\frac{1}{2}$p

3. Multiply each of the following by 100. (Write the answers in £.)

   a. 17p b. 31p c. 2$\frac{1}{2}$p d. £2·25 e. £0·14 f. £0·04$\frac{1}{2}$
   g. 12$\frac{1}{2}$p

4. Use the rounding off method to find the cost of:

   a. 21 articles at 3p each b. 19 articles at 8p each
   c. 23 articles at 7$\frac{1}{2}$p each d. 18 articles at 6$\frac{1}{2}$p each
   e. 43 articles at 5$\frac{1}{2}$p each f. 99 articles at 7p each

5. A shopkeeper has to work out the following bills. Complete them for him.

   a. $\frac{1}{2}$ litre oil @ 20p per litre
   20 large eggs @ 2$\frac{1}{2}$p each
   $\frac{3}{4}$ kg meat @ 48p per kg
   2 litres milk @ 15$\frac{1}{2}$p per litre ____
   Total ____

   b. 38 stamps @ 2$\frac{1}{2}$p each
   43 stamps @ 2p each
   3 postal orders @ 12$\frac{1}{2}$p each
   Postage on parcel £0·42 ____
   Total ____

6. Paint can be bought in 1-litre, $\frac{1}{2}$-litre, and $\frac{1}{4}$-litre tins.

   a. How much do I save if I buy a 1-litre tin instead of two $\frac{1}{2}$-litre tins?

   b. How much do I save if I buy a $\frac{1}{2}$-litre tin instead of two $\frac{1}{4}$-litre tins?

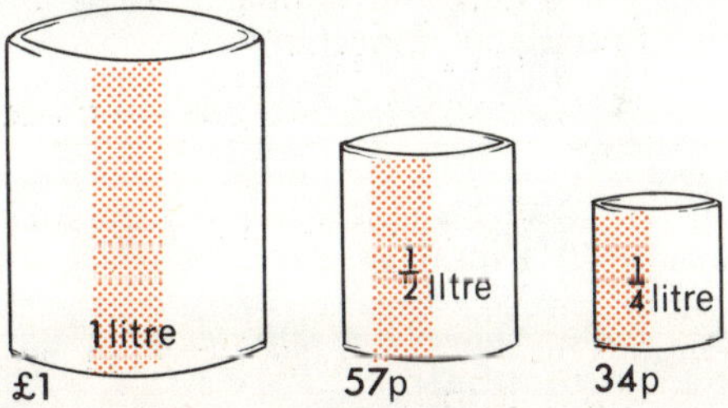

7. Use any method you wish to work out:

   a. £12·25×50 b. £0·12$\frac{1}{2}$×24 c. 2$\frac{1}{2}$p×64
   d. £12÷25 e. £4·48÷8 f. £6÷12$\frac{1}{2}$

Let's Discover **Curved Surfaces**

1. a. Cut out a circle of card with a radius of 4 cm and draw a diameter on it. Attach a rubber band near each end of the diameter.

   b. Stretch your rubber bands until they are tight. Ask your neighbour to "wind up" the circle and then let it spin.

   c. Did you get the impression of a ball as the card was spinning?

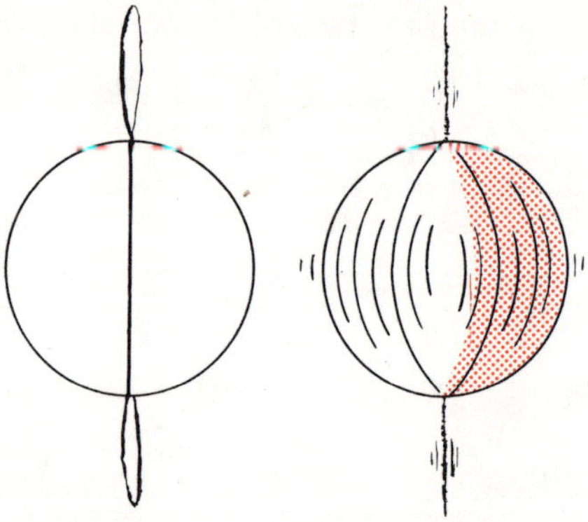

- The mathematical name for a ball is a **sphere**.

The centre of your card is called the centre of the sphere.

Here are some pictures of things which are spheres.

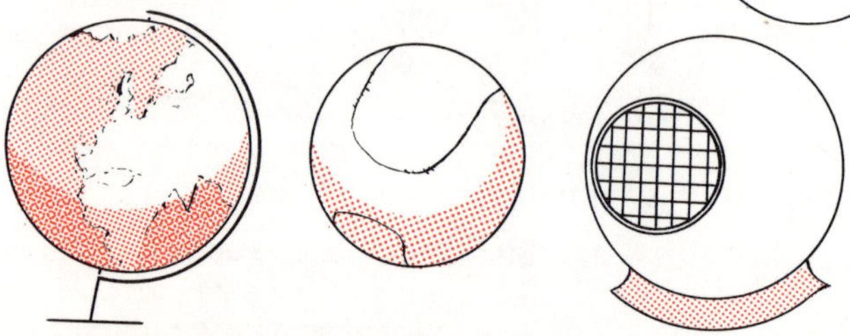

2. Cut out two circles of card with a radius of 4 cm.
   Cut along one radius of each and fit the two pieces together.
   What shape are you beginning to build?

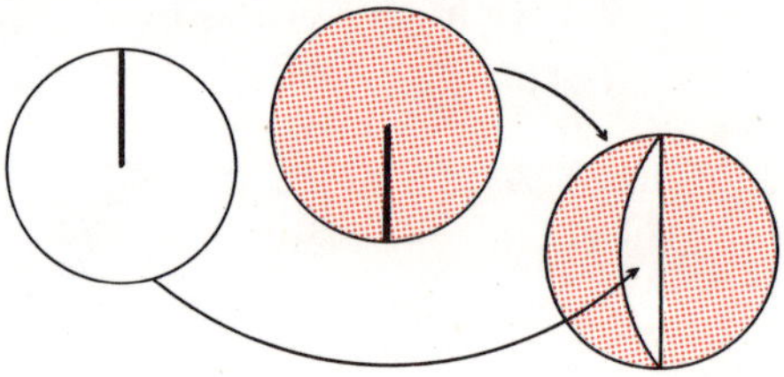

3. a. Make a sphere out of plasticine and cut through it.

- Each half of the sphere is called a **hemisphere**.

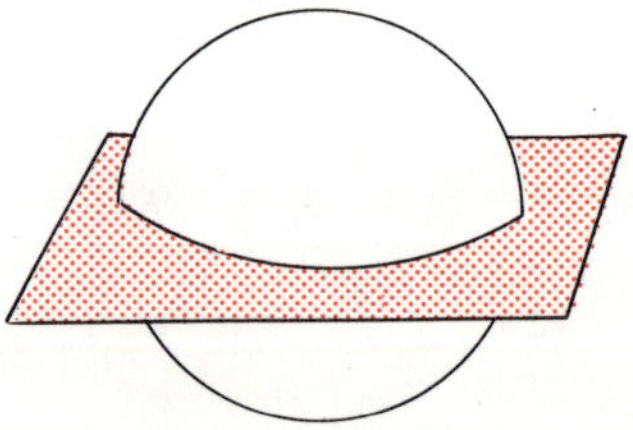

   b. Which element from the set of points {A, B, C, D} was inside the sphere?

   c. Which elements of the set were on the sphere?

   d. Is the distance of point B from the centre of the sphere the same as the distance of point C from the centre of the sphere?

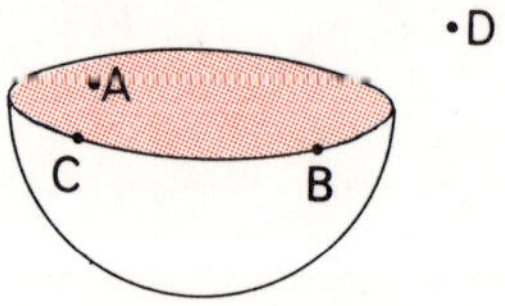

- The distance of each point on the sphere from the centre of the sphere is called the **radius**.

Let's Try

1. Choose a sphere which you think has a radius of about 3 cm. Can you measure its radius?
   Try using a pair of calipers. The distance AB will be 2×**radius** of the sphere.

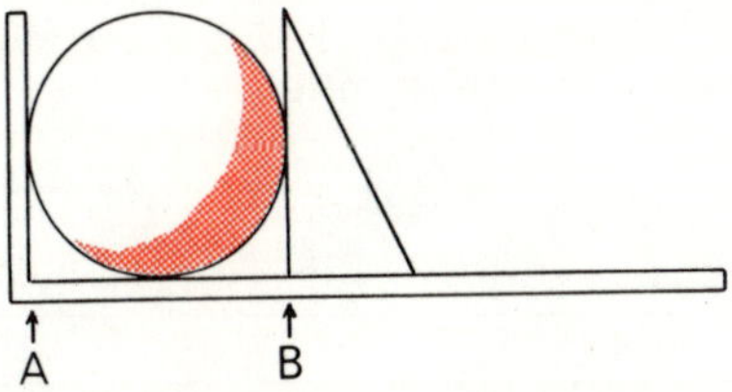

2. a. Make a collection of spheres and hemispheres.
   b. Measure the radius of each sphere and put them in order of size.

3. From the set {A, B, C, D, E}:

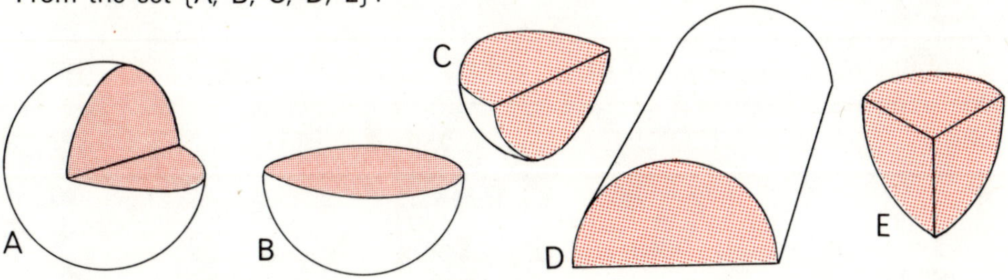

   a. Which element is not part of a sphere?
   b. Which element is a hemisphere?
   c. Which element used twice would make a hemisphere?
   d. How many pieces like C are needed to make a sphere?
   e. How many pieces like E are needed to make a sphere?

4. If you cut through a sphere, what curve do you make?

5. Can you cut through an orange and not make a circle?

6. How are these two shapes different?
   Think about the way in which they roll.

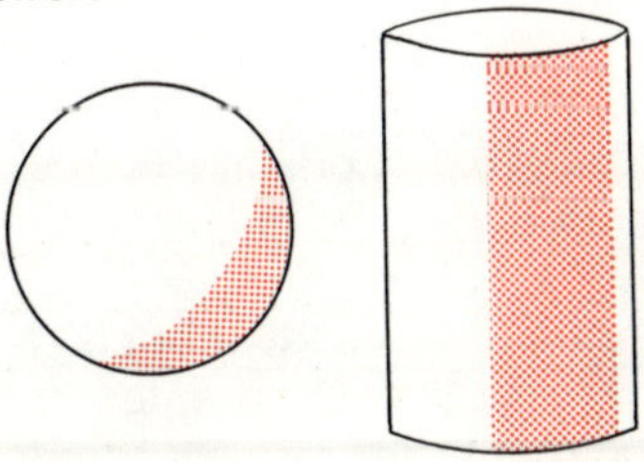

7. If wheels were shaped like spheres through which point would the axle have to pass?

Let's Discover

Weights from Graphs

Aniseed balls are popular sweets because they last a long time.

I counted the number of aniseed balls I received in a 100-g bag.
There were 80 balls in the bag.

1. How many aniseed balls should weigh:

   a. 200 g b. 300 g c. 1000 g

   Since a 100-g bag contains 80 aniseed balls, a 1-kg bag contains 80×10 = 800 balls

   Instead of making a lot of calculations we could draw a graph like the one opposite.

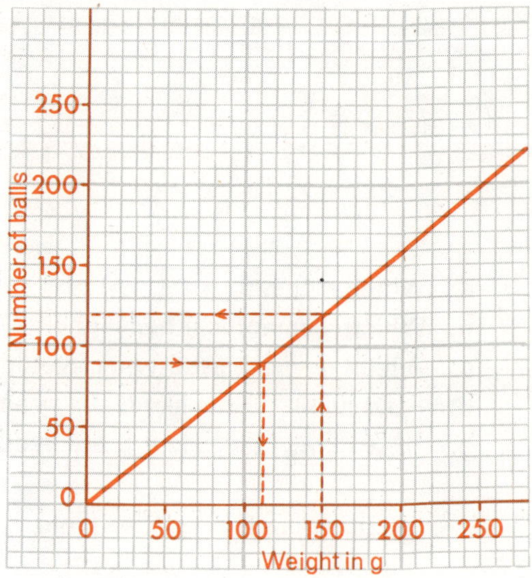

2. Copy the graph onto 2-mm graph paper.
   Extend the axes to 1000 g and 800 balls.

   For example, in order to find out how many balls weigh 150 g we can follow up ↑ from 150 g and across ← to 120.
   120 balls weigh 150 g

   What is the weight of 90 balls?
   In order to answer this question, follow across → from 90 and down ↓ to about 110 g.

3. Do 90 balls weigh more or less than 110 g? Look at the graph carefully.

   Mother made some chocolate fudge for the school fête.
   She found that a 5-cm square weighed 80 g.

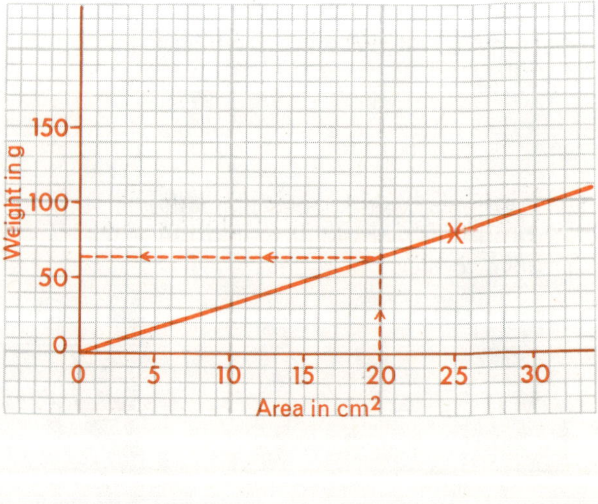

4. Complete this working:
   Area of a 5-cm square = 5×5 = . . . cm².

   Note that a x has been shown at (25, 80) on the graph.

A piece of fudge 4 cm wide and 5 cm long has an area of 20 cm^2.

5. By following up ↑ from 20 cm^2 and across ←, what is the approximate weight of this piece?

Let's Try

1. Use your graph showing the weight of aniseed balls to check how many balls weigh:

   a. $\frac{1}{2}$ kg b. 400 g c. 250 g d. 750 g

2. Use your graph showing the weight of aniseed balls to find the approximate weight of:

   a. 250 b. 700 c. 500 d. 350 aniseed balls

3. Complete the graph showing the weight of chocolate fudge. Draw axes 0 to 60 cm^2 and 0 to 200 g.

   a. What area will have a weight of:
   100 g 1000 g

   b. What is the weight of a piece:
   2 cm × 15 cm; 1 cm × 30 cm
   Which of these two pieces would you prefer to eat?

4. Here is a graph showing my pet's weight.

   a. What length represents 1 kg on the weight axis?

   b. Write the weight of the animal in kg when he is:
   1 month; 2 months; 5 months old

   c. Write his weight in g when he is:
   1 month; 2 months; 5 months old

   d. Write his weight at birth.

   e. At what age did he weigh 2·3 kg?

   f. At what age did he weigh 2·7 kg?

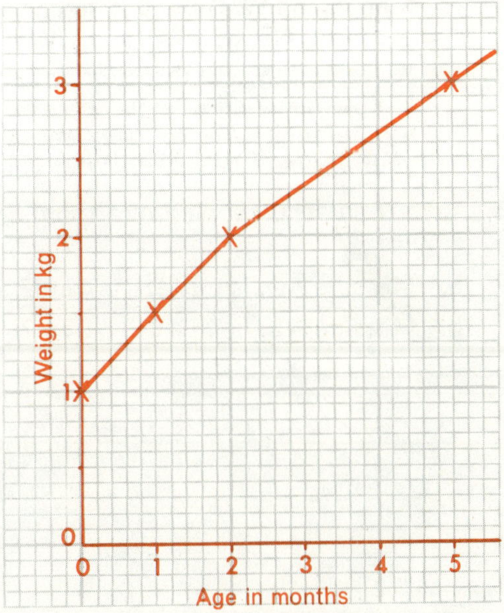

Let's Discover — Compass Bearings and Scales

1. a. Cut out a circular piece of paper and fold it into quarters.
Open it up and mark it as in this diagram.

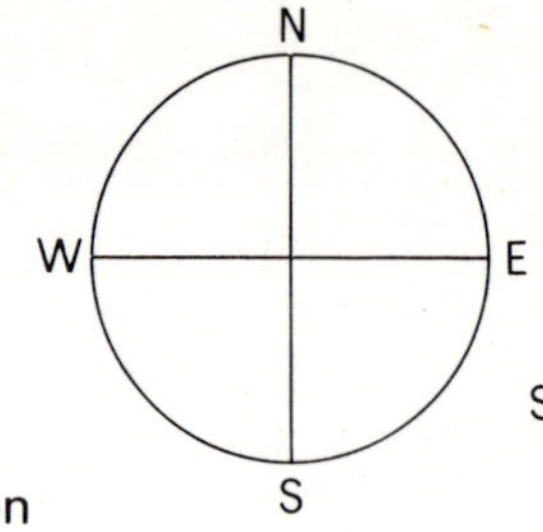

N stands for north.
E stands for east.
S stands for south.
W stands for west.

b. Have you noticed weather vanes on churches and other buildings?
Find out how a weather vane works.
Perhaps you could make one.

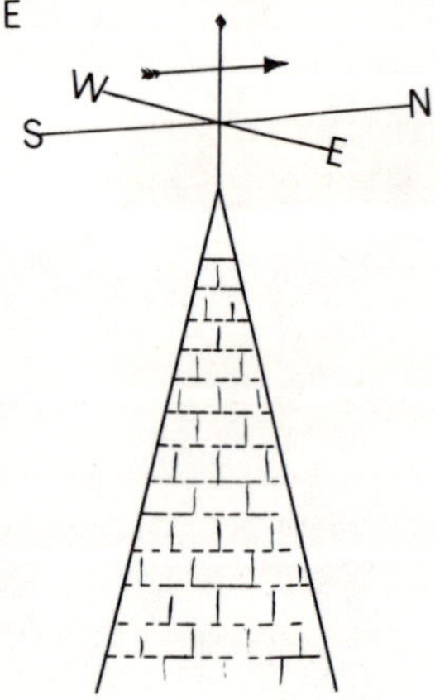

This is called a **compass**. The needle always points to magnetic north (N).

2. a. Lay your paper circle on your desk so that the arrow marked N is pointing to magnetic north. (Use a compass to help you locate N.)

b. Point your arm south, then east, then west.

3. Fold your paper circle up again, and make another fold which will divide it into 8 equal parts. Put the extra compass points on it.

NE is north-east. NW is north-west.
SE is south-east. SW is south-west.

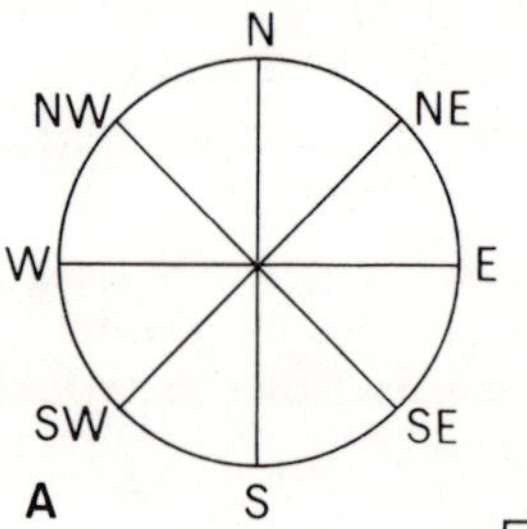

A

4. a. Copy diagram B onto squared paper. Mark E and W. Draw line segments and mark NE, SE, SW, and NW.

b. If you face N and then turn to face E, how many right angles do you turn through? Are you sure about your answer? Which way did you turn?

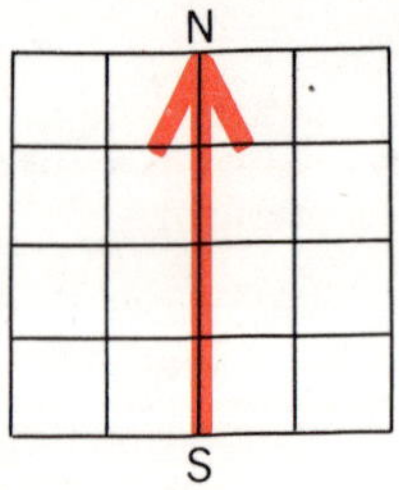

B

5. Here is an accurate plan of a room in a house.

a. Measure $\overline{AD}$. Is $m(\overline{AD}) = 4$ cm?

b. Measure $\overline{AB}$. Is $m(\overline{AB}) = 5$ cm?

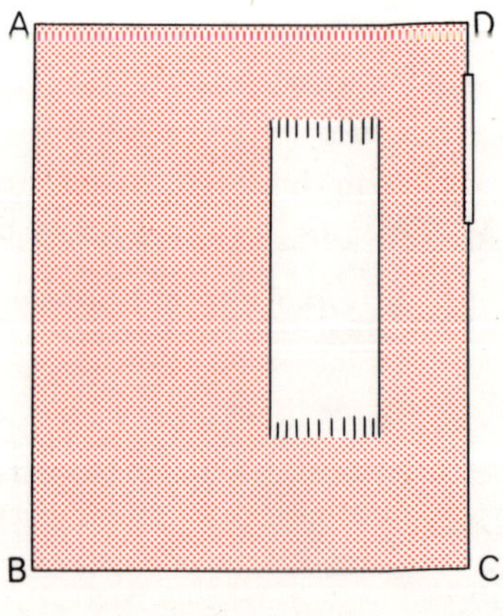

- In order to draw plans and maps we have to let each distance on land be represented by a line segment on paper. We say that the map or plan has been **drawn to scale.**

The scale used on this plan is 1 cm represents 1 metre, or **scale**: 1 cm to 1 m.

6. Since the room is 4 cm wide on the plan it is actually 4 metres wide. What is the actual length of the room?

Let's Try

1. These pictures have been drawn to scale. Copy and fill in this table after measuring each article.

Scale: 1 cm to 1 m	**Length**	**Real Length**
	(a) . . . cm	
	(b) . . . cm	
	(c) . . . cm	
	(d) . . . cm	

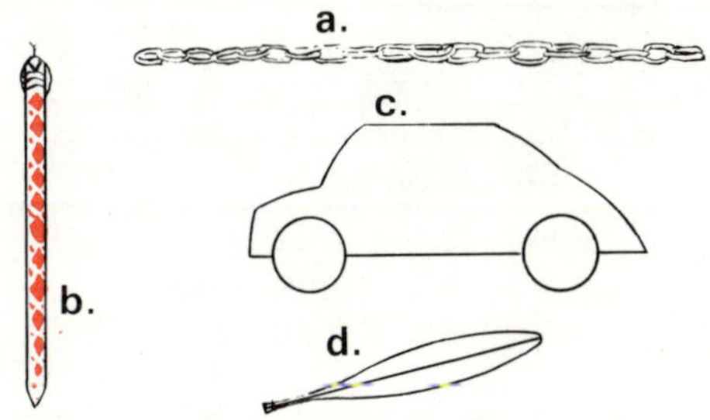

2. Using a scale of 1 cm representing 1 m draw line segments which represent:

a. 12 m **b.** 8 m **c.** 5 m **d.** 50 cm **e.** 9·5 m

3. These line segments have been drawn using a scale of 1 cm representing 1 km. Measure each segment and write the distance which it represents.

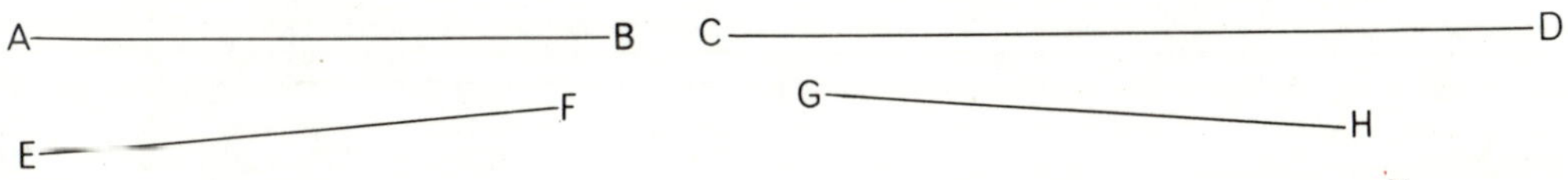

4. The scale used on this map is 1 cm represents 1 km.

a. Use a ruler to measure $\overline{AB}$ and $\overline{BC}$ in cm.

b. How far is it from A-town to B-town, and B-town to C-town?

c. If you stand at B-town and face C-town, in which direction are you facing?

d. What is the direction of C-town **from** B-town, and B-town **from** A-town?

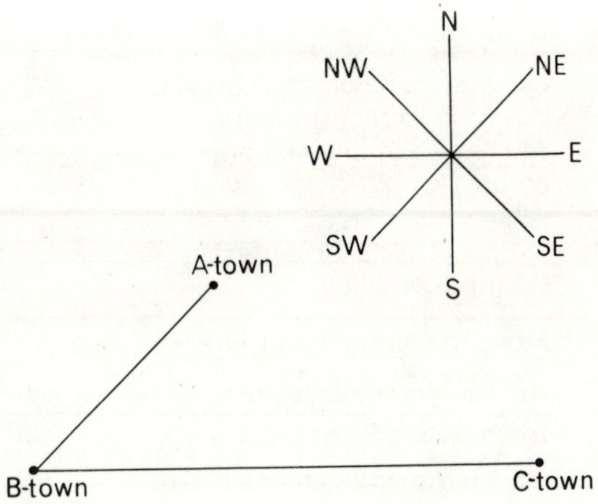

Let's Discover

More Surfaces

1. a. Cut two regular pieces of card 20 cm by 5 cm like these and cut them on the black lines.
   b. Think of the cards as planes and fit them together to make a cross.
   c. Can you see the line segment formed by the intersection?

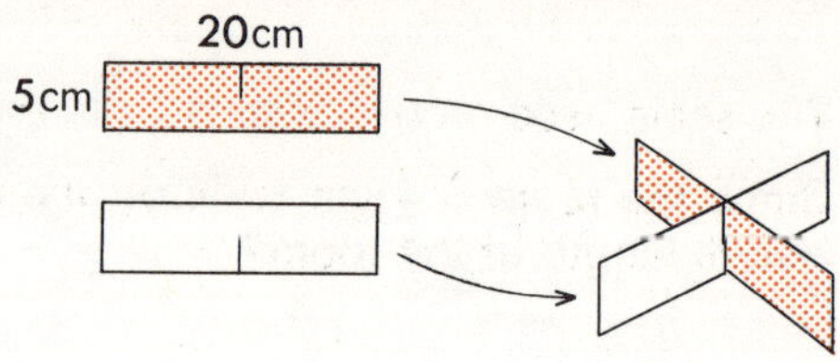

This is a picture of a doll's house. Think of the walls, floors, and ceilings as planes, although they are really plane segments.

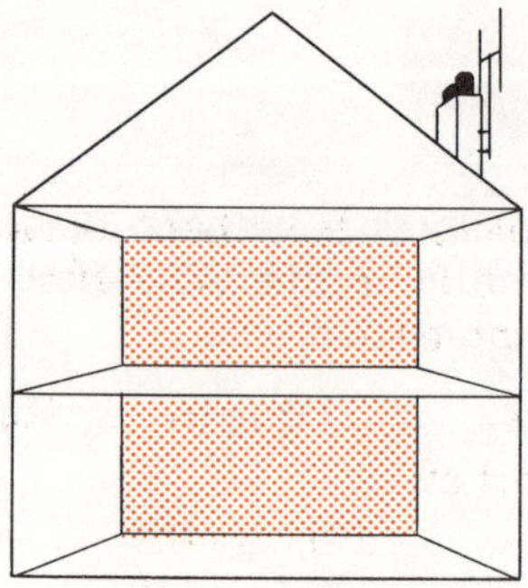

2. a. How many planes can you see?
   b. How can you describe the intersection of:
      a wall and a floor?
      a wall with another wall?
      a wall and a ceiling?

Stand your cross on a table-top which you can think of as a third plane, C.

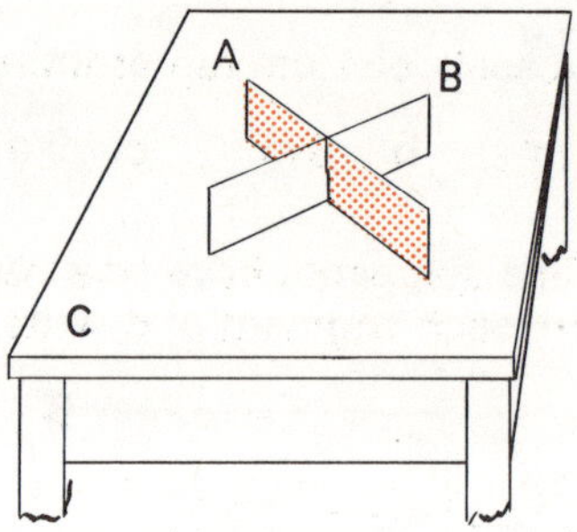

3. a. Planes A and C intersect in a . . .
   b. Planes B and C intersect in a . . .
   c. Planes A, B, and C intersect in a . . .

- On this cuboid you can see that each plane surface is called a **face**.
- Where two faces intersect in a line segment, the line segment is called an **edge**.
- Where three faces intersect in a point, the point is called a **vertex**. (Plural – vertices)

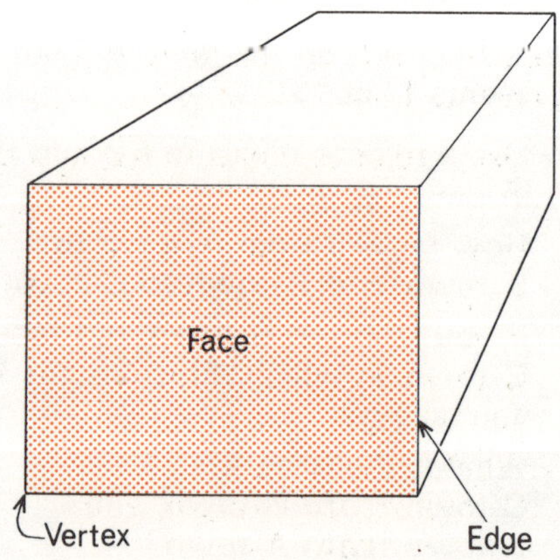

4. From this cuboid can you say:
   a. how many faces it has?
   b. how many edges it has?
   c. how many vertices it has?

Let's Try

1. Collect a set of solids such as cubes, spheres, . . .

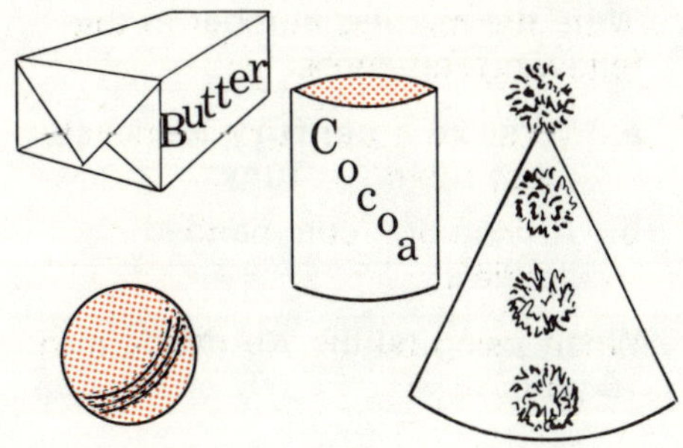

   a. Which of your solids have no vertices?
   b. Which of your solids have plane faces?
   c. Which of your solids will roll?
   d. Make and describe as many subsets as you can from your set.

2. Here are some pages from a book. Make a model like this and think of each sheet of paper as a plane.

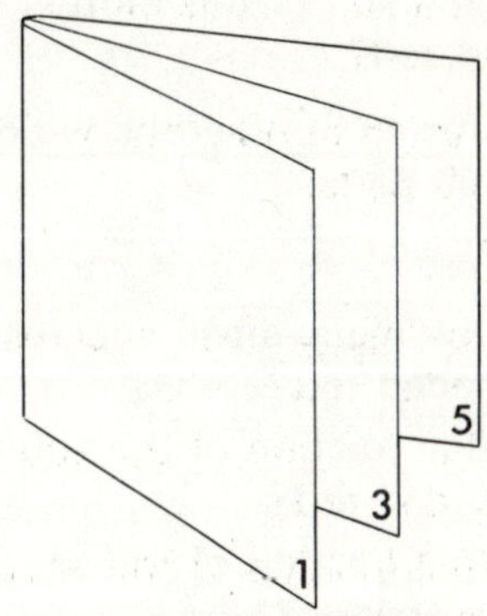

   a. How many planes have you?
   b. What do we call the intersection of pages 1 and 3?
   c. What do we call the intersection of pages 1, 3, and 5?
   d. Can you open your pages so that you seem to have only two planes?

3. This is a plan for a cycle shed. The plane segment for the door is named by ABFE.

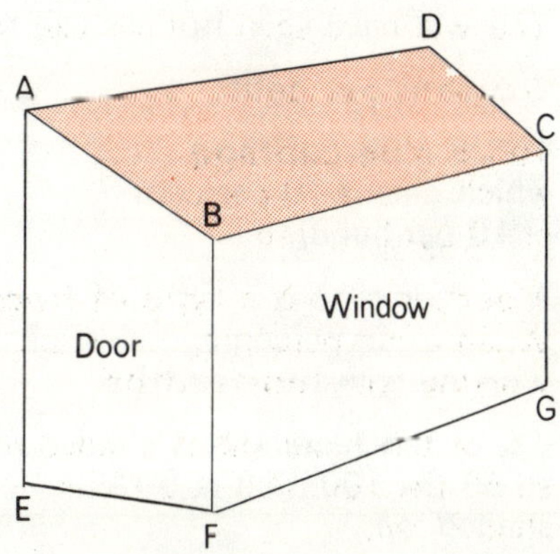

   a. Name the plane segment which intersects with ABFE in the line segment $\overline{BF}$.
   b. Name the two plane segments which intersect in $\overline{AB}$.
   c. Name the line segment which is the intersection of plane segments ABCD and FGCB.
   d. How many faces can you see?
   e. How many edges can you see?
   f. How many vertices can you see?

4. This is a 3-cm cube.

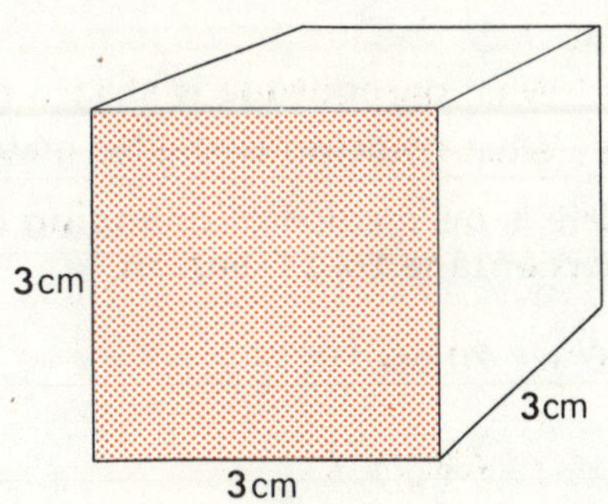

   a. What is the total length of all the edges?
   b. Work out the total area of all the faces in cm^2.

Let's Discover **Percentages**

1. Write the missing number in the following sentences:

a. To score a **century** a cricketer must score . . . runs.

b. A centurion commanded . . . soldiers.

When using tenths we divide into 10 parts.

2. What fraction of this strip is shaded red?

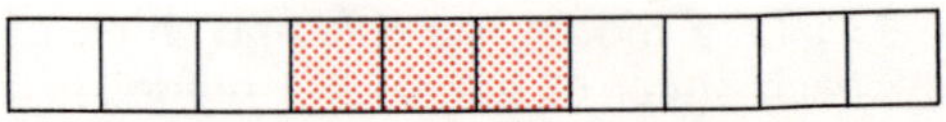

When using hundredths we divide into 100 parts.

3. **a.** How many small squares are shaded red?

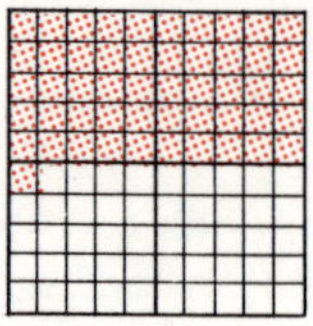

b. What fraction of the square is shaded red?

c. What fraction of the square is not shaded?

You will have seen notices like this.

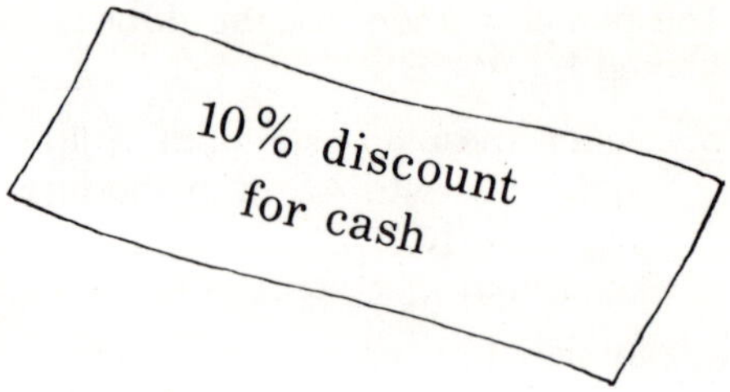

- **% means per cent.**

10% is a **percentage** which means 10 per cent or 10 per hundred.

- A **percentage** is a **type of fraction.** When using percentages we are working with **hundredths.**

$\frac{50}{100}$ of this rectangle is shaded red, so 50 per 100, that is 50%, is shaded red.

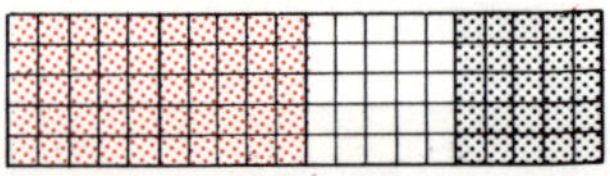

$50\% = \frac{50}{100} = \frac{1}{2}$

4. **a.** What **percentage** is shaded grey?

b. What **fraction** is shaded grey?

Here is the method of **changing a percentage to a fraction.**

20% = 20 per 100

$= \frac{\cancel{20}^{1}}{\cancel{100}_{5}} = \frac{1}{5}$

5. Complete the following working:

a. $\frac{1}{10}$ is 1 out of 10 which is 10 out of . . . so $\frac{1}{10}$ = 10%

b. $\frac{2}{5}$ is 2 out of 5 which is 4 out of . . . which is 40 out of 100 so $\frac{2}{5}$ = . . . %

• To change a fraction to a % we must change it to an equivalent fraction with 100 as the denominator.

$\frac{3}{5} = \frac{6}{10} = \frac{60}{100}$, so $\frac{3}{5}$ = 60%

Let's Try

1. Make a list of all the places where you can see %. Collect posters and pictures with a percentage sign on them.

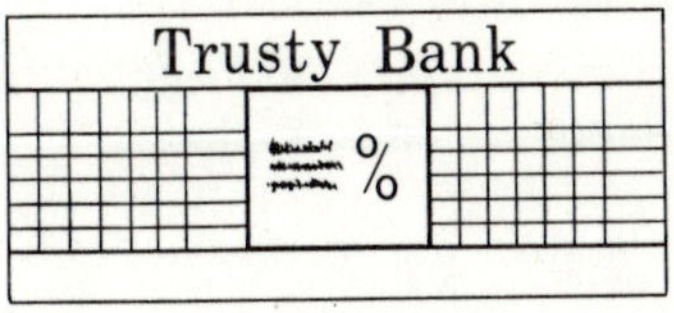

2. Write the following as percentages:

a. 1 out of 100 b. 10 out of 100 c. 5 out of 100
d. 50 out of 100 e. 15 out of 100 f. 100 out of 100

3. Write these %s as fractions and then as decimals:

a. 20% b. 40% c. 25% d. 75%

4. Change these fractions to %s:

a. $\frac{1}{20}$ b. $\frac{3}{20}$ c. $\frac{4}{5}$ d. $\frac{3}{4}$

5. Here is a graph showing pets kept by Class 3.

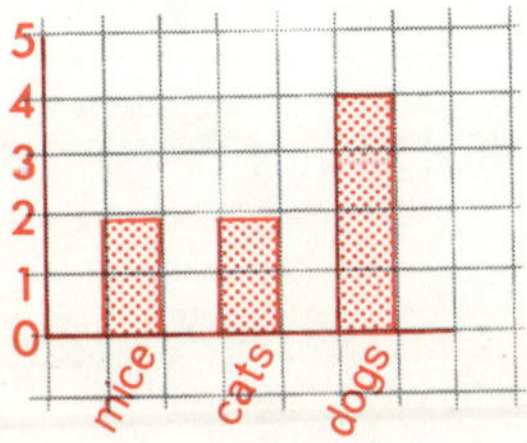

a. What fraction is dogs? What percentage is dogs?

b. What fraction is cats? What percentage is cats?

6. The rate of interest in a Post Office Savings Account is $2\frac{1}{2}$%. Find out:

a. what bank interest is.

b. how much interest you get on £100.

Let's Discover **The Set of Integers**

1. Can you give away 10 marbles if you have only 8 marbles?

   There is no replacement for △ in
   8−10 = △ if △ is a **natural number**.
   {1, 2, 3, 4, . . .}

- The set of natural numbers is **not closed** for **subtraction** since the difference between two of them is not always a natural number.

2. Here is an astronaut waiting for **zero hour**.
   6 seconds before launching is "Z minus 6".
   2 seconds after launching is "Z plus 2".

   a. What do we call:

      3 seconds before Z hour?
      1 second after Z hour?

   b. How many seconds are there between "Z minus 5" and "Z plus 3"?

   The astronaut's count-down clock looks like this.

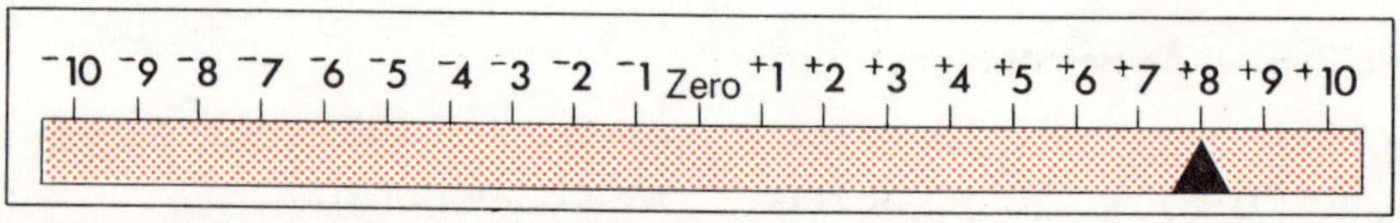

This count-down clock shows **plus** $^{+}$ to the **right** of **zero** and **minus** $^{-}$ to the **left** of **zero**.

3. Use the count-down clock to help you complete this sentence.
   "10 seconds before $^{+}8$ is . . . so △ − $^{+}10$ = $^{-}2$.

   The astronaut's clock is like this number line.

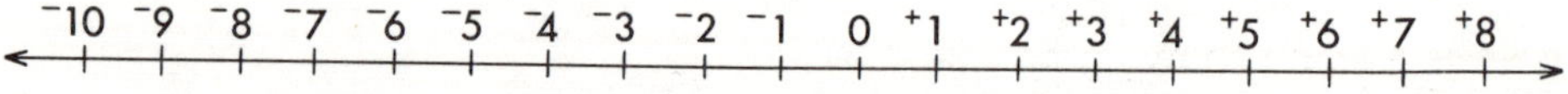

   This number line shows the **set of integers**.
   {. . . $^{-}6$, $^{-}5$, $^{-}4$, $^{-}3$, $^{-}2$, $^{-}1$, 0, $^{+}1$, $^{+}2$, $^{+}3$, $^{+}4$, $^{+}5$, . . .}

- The set of integers contains **positive** numbers, **negative** numbers, and **zero** or "nought".

   $^{+}6$ is a positive number and reads "**positive six**".
   $^{-}7$ is a negative number and reads "**negative seven**".

4. On this number line the set {$^{-}6$, $^{-}3$, 0, $^{+}3$, $^{+}6$} has been graphed with ×.
   Write the set which has been graphed with •.

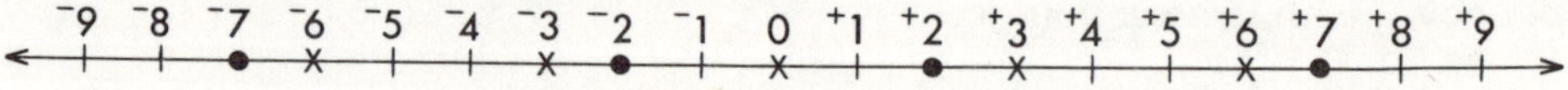

From the **integer number line**, sentences such as "$^+4 - ^+6 = \triangle$" can be completed.

$^+4 - ^+6 = ^-2$

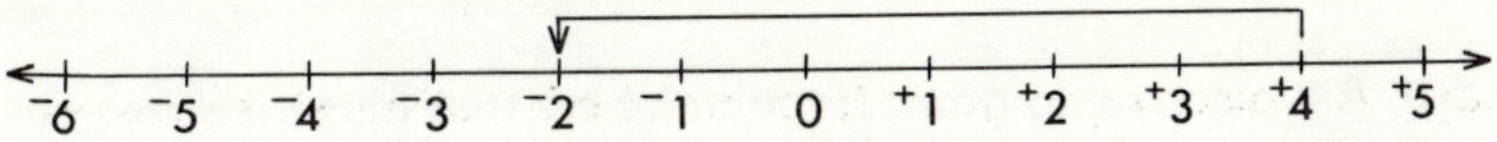

Remember $^-$ means **move left** and $^+$ means **move right** on the line.

5. Use the number line to find the number which replaces $\triangle$ in:

a. $^+5 - ^+3 = \triangle$ **b.** $^+4 - ^+7 = \triangle$ **c.** $^-3 + \triangle = ^+4$

Let's Try

1. Draw a number line like this one:

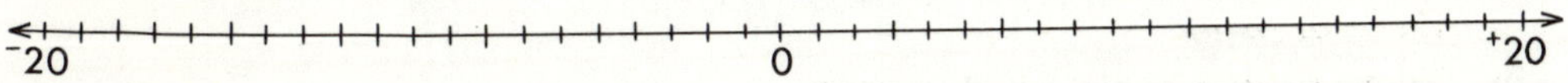

a. Graph A = $\{^-20, ^-15, ^-10, ^-5, 0, ^+5, ^+10, ^+15, ^+20\}$ with ×.

b. Graph B = $\{^-18, ^-15, ^-12, ^-9, ^-6, ^-3, 0, ^+3, ^+6, ^+9, ^+12, ^+15, ^+18\}$ with •.

c. Write $A \cap B = \{\ldots\}$.

d. Which numbers are marked both × and • ?

2. Use the number line shown in question **1** to complete:

a. $^+4 + ^+7 = \triangle$ **b.** $^+4 - ^+7 = \triangle$ **c.** $\triangle + ^+8 = ^+15$

d. $\triangle - ^+8 = ^-1$ **e.** $^+8 - \triangle = ^-3$ **f.** $\triangle + ^+10 = ^+5$

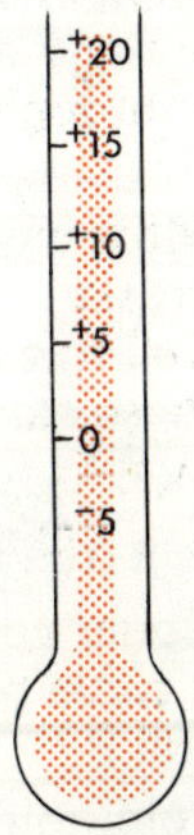

3. Here is a thermometer. Water freezes at 0 °Celsius.

a. The temperature is 20 °C during the day-time. Will water freeze if the temperature falls:
(i) 15° at night? (ii) 22° at night?

b. The temperature at night was $^-4$ °C. What will the temperature be if it rises 8°?

c. Complete this table.

Temperature (start)	$^+5$	$^+5$	$^+5$		$^-5$	degrees
Rise $^+$ or Fall $^-$	$^+5$	$^-5$		$^-5$	$^+10$	degrees
Temperature			$^-5$	$^-3$		degrees

Maps and Plans

Klondike Bill discovered gold. He remembers that his strike was about $1\frac{1}{2}$ km from Blackville.

1. **a.** Copy this map and use a pair of compasses to locate on the map all the points at which the gold could be.

b. Do the set of points at which the gold could be lie on a circle?

c. Bill now remembers that he headed SE from Blackville on the day he discovered gold. Mark with G the point at which the gold is most likely to be.

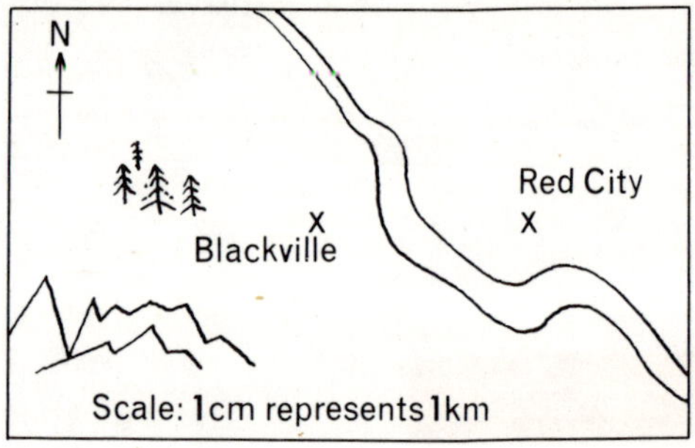

2. **a.** Mark two points A and B on a piece of paper.

b. Try to draw a circle which passes through the two points. I missed one of the points with each of my first two tries.

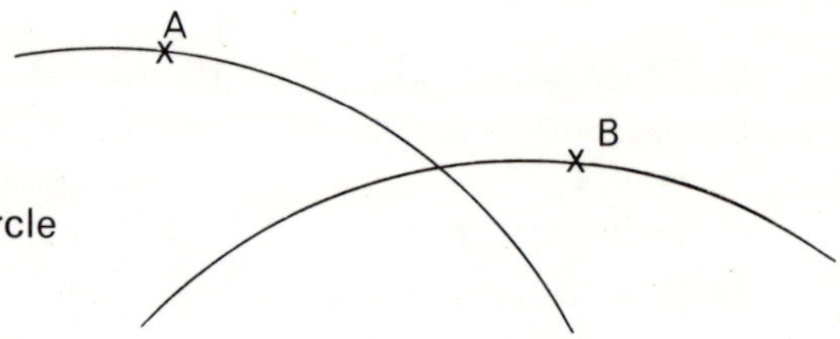

Here is a simple way of drawing a circle through two fixed points. Carefully put a pin through both points and then crease your paper.

3. Put your compass point anywhere on the crease, then open the legs until the pencil is resting on point A. Now draw a circle. Does your circle pass through B?

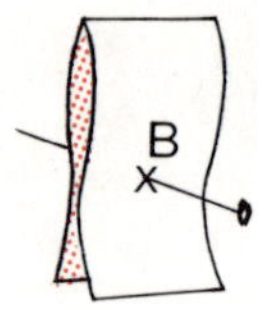

4. **a.** Mark any point you like on the crease. Call this point C. Complete these statements:
$m(\overline{CA}) = \ldots$ $m(\overline{CB}) = \ldots$

b. Is it true that $m(\overline{CA}) = m(\overline{CB})$?

c. Could you use C as a centre and draw a circle passing through points A and B?

d. Join up points A and B with a line segment AB.

e. Do $\overline{AB}$ and the crease cross at right angles?

f. Complete this statement:
The crease divides $\overline{AB}$ into two . . . parts.

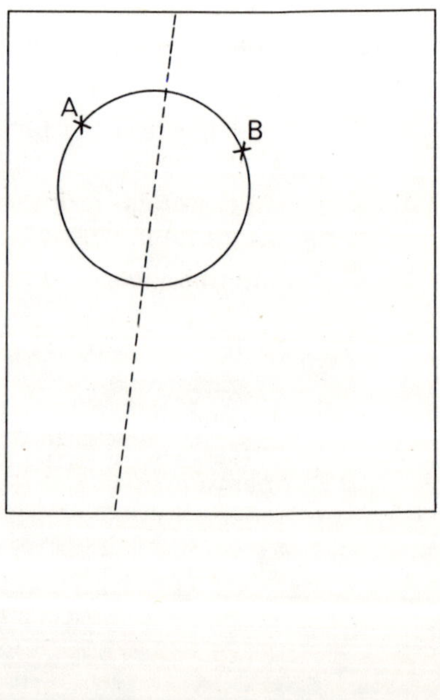

Klondike Bill now remembers that his gold was at a place which was as far from Blackville as it was from Red City.

5. a. Copy this map using a scale of 1 cm represents 1 km. Blackville (B) is 2 km from Red City (R).
   b. Use a ruler to locate and mark some points on your map which are the same distance from B as from R.
   c. Join B to R and mark the middle point of BR. Fold your piece of paper so that the × at R falls on the × at B.
   d. Where could the gold lie?

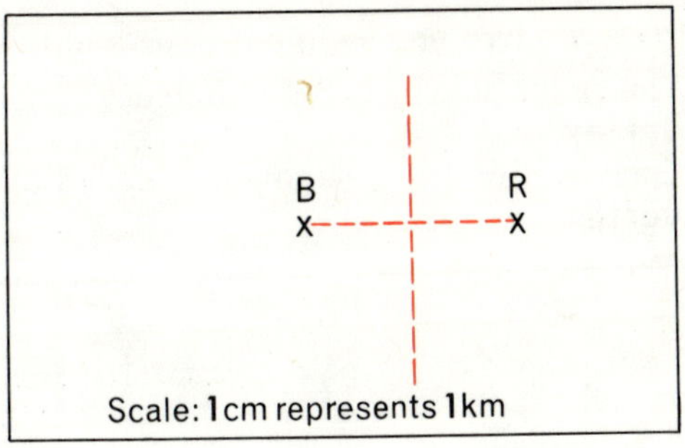

Let's Try

1. The scale on this map is 1 cm represents 1 km.
   a. How long are the segments $\overline{AB}$, $\overline{BC}$, and $\overline{AC}$? How many km does each segment represent?
   b. In what direction would you travel if you were flying from A to B; B to A; and C to B?

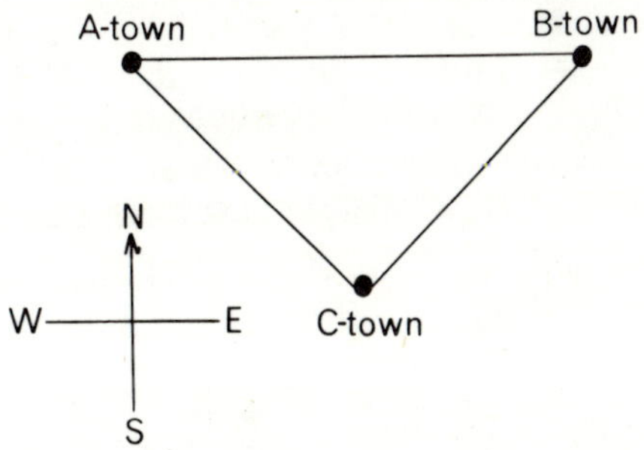

2. Mark two points A and B on a piece of paper. Make $m(\overline{AB}) = 8$ cm.
   a. Point C is 5 cm from A. Draw a circle which shows all the possible positions of C.
   b. Point C is 5 cm from B. Draw a circle which shows all the possible positions of C.
   c. At how many points on your paper could you put C?

3. Here is a map which has been drawn to scale.
   a. If it is 4 km from Squareton to Roundton, how far is it from Roundton to Pointon?
   b. Draw the map accurately.
   c. Can you locate the old oak tree which is 4 km from Roundton and is as near to Pointon as it is to Roundton?

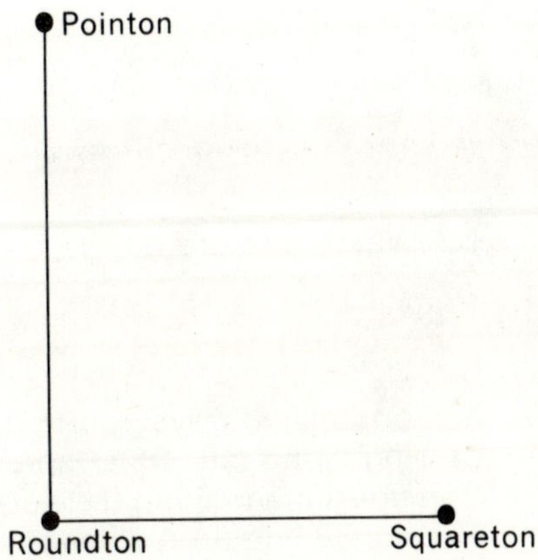

Let's Discover **Networks**

Here are two line segments, $\overline{AB}$ and $\overline{PQ}$.
The point at which they intersect is called a **vertex**.
Each end of a line segment is called a **vertex**.

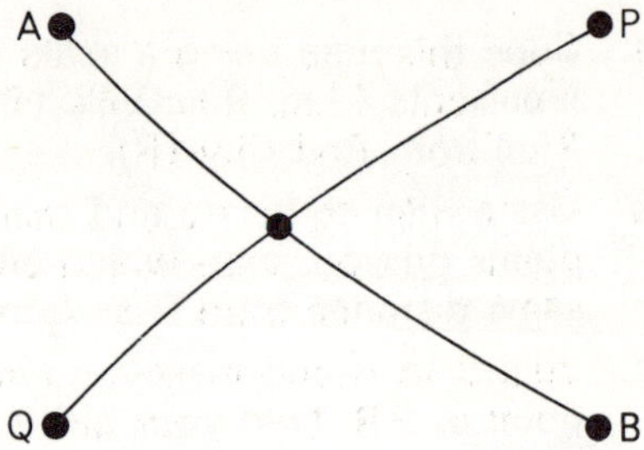

1. a. Which two of the following are **not closed curves**?

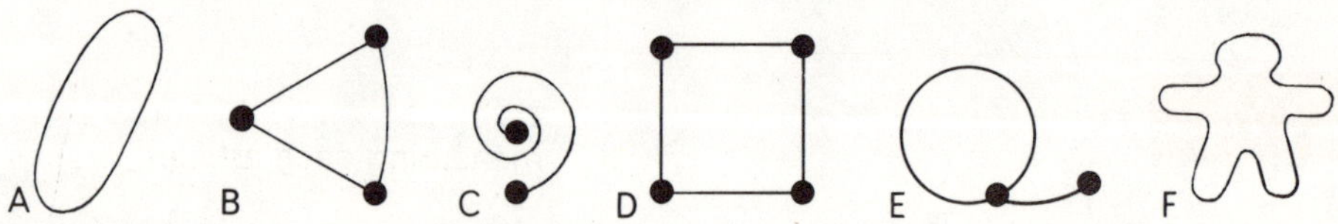

b. Curve D has four vertices, ●. Curve C has two vertices, ●.
How many vertices have curve B, and curve F?

c. Curve D has 4 edges. Curve E has 2 edges.
How many edges have curve B, curve F, and curve C?

Here is another set of curves. The curves are all closed.

Curve H divides the plane of this page into three regions: the orange and the grey shaded regions, and an outside region.

2. a. Into how many regions is the plane divided by I; J; K; and L?

- A **simple closed curve** divides the plane into only two regions.

b. Which one of H, I, J, K, and L is a **simple closed curve**?

- Diagrams like H, I, J, K, and L are called **networks**.

3. Is it possible to travel round the networks at the top of the next page without passing along any edge more than once?
Run your finger along the edges of each network until you find the right path.
The dotted line on A shows one way to start.

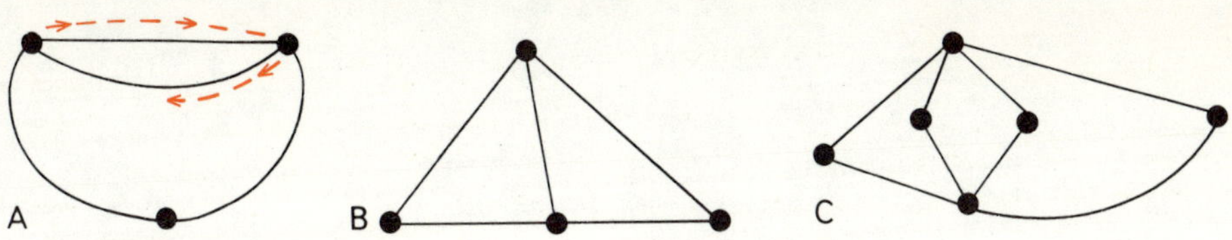

Let's Try

1. Complete this table for the five networks.
If you can travel round the network without passing along any edge more than once put **yes** in the empty box. If you can't, put **no** in the box.

Network

A	
B	
C	
D	
E	

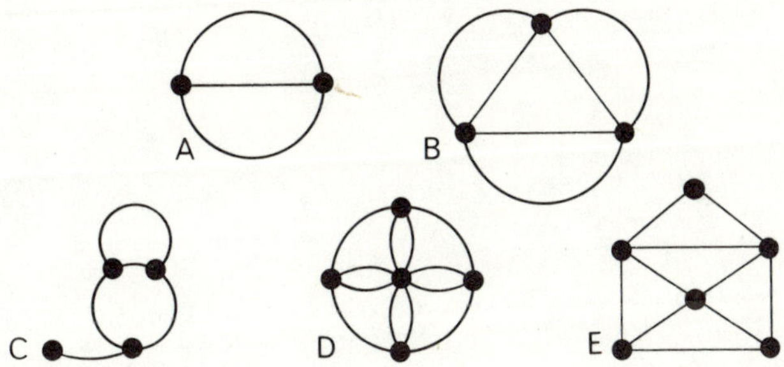

2.

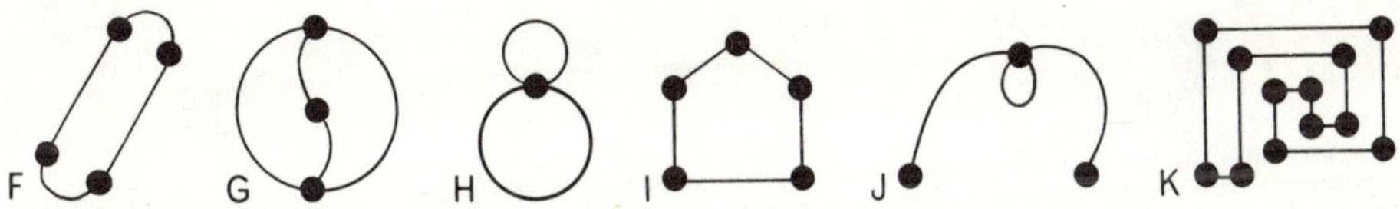

a. Which one of this set of networks is not a closed curve?

b. Which of the networks are simple closed curves?

c. Copy and complete this table.

Network	F	G	H	I	J	K
Number of vertices (V)		3				
Number of edges (E)	4		2			
Number of regions plane is divided into (R)					2	

d. Complete this relationship between V, E, and R? V−E+R = . . .

Answers

Answers

Page 7

1. **a.** 16 **b.** 8
2. 6 tiles

Page 9

1. **a.** 5 **b.** 3 **c.** 3 **d.** 4 **e.** 7 **f.** 4 **g.** 4 **h.** 10

2. **a.**

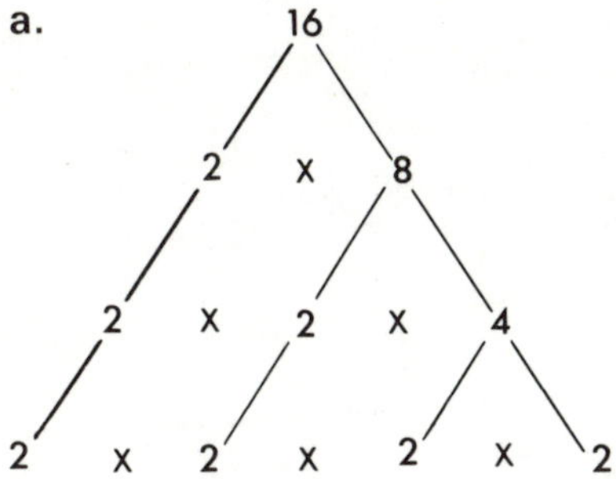

b.

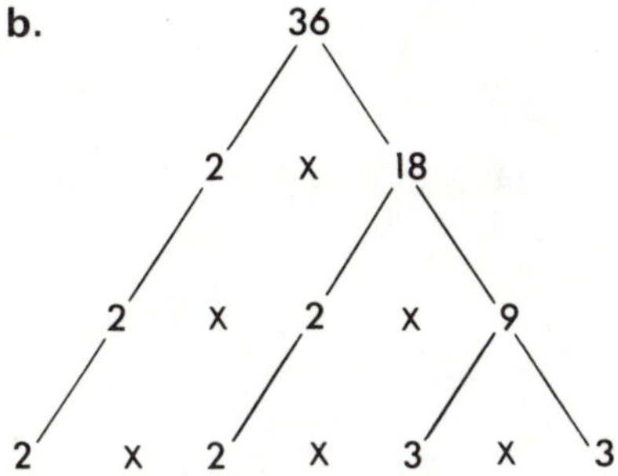

c. {1, 2, 8, 4, 16} **d.** {1, 2, 18, 9, 3, 4, 6, 12, 36}

3. **a.** {1, 3, 5, 15} **b.** {1, 2, 4, 5, 10, 20} **c.** {1, 2, 3, 6, 9, 18}
   d. {1, 3, 5, 9, 15, 45} **e.** {1, 3, 11, 33} **f.** {1, 23}

4. **a.**

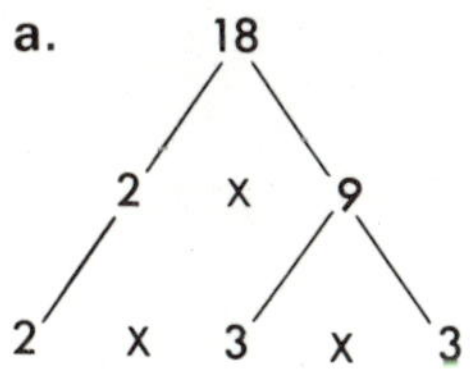

b.

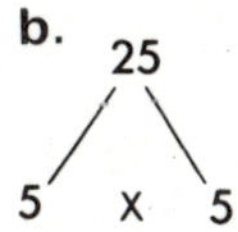

c.

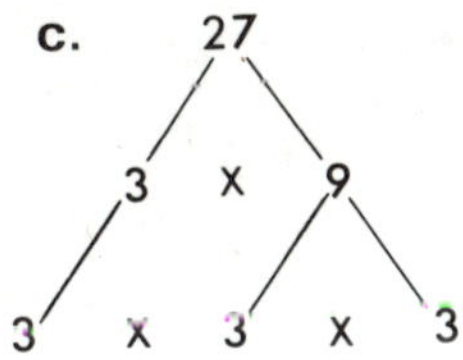

d.

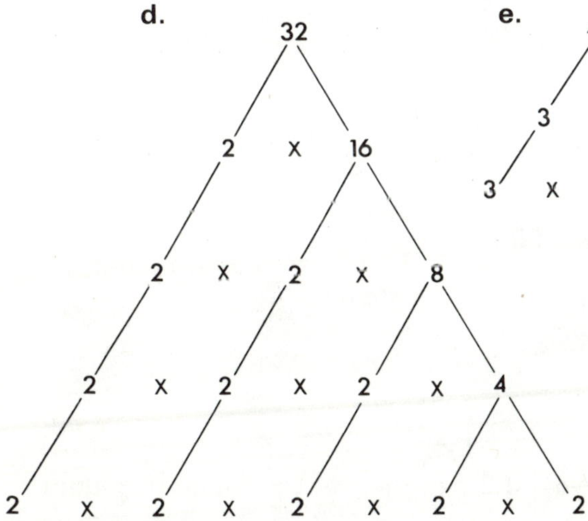

e.

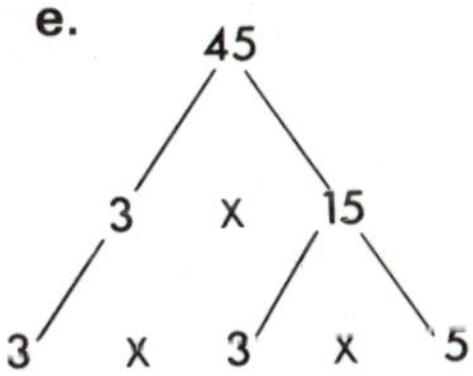

f.

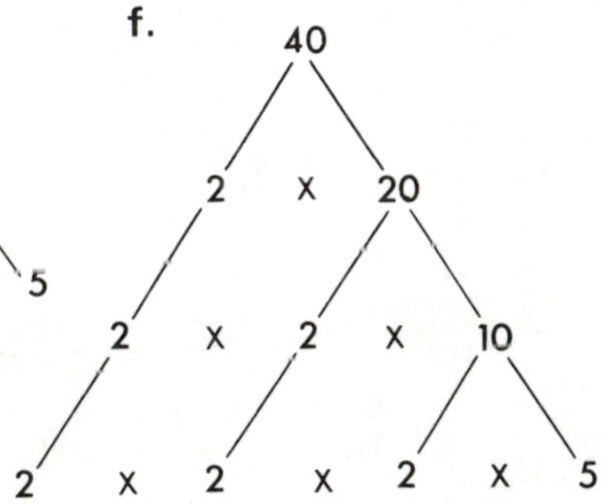

5. **a.** 1064 **b.** 1560 **c.** 2272 **d.** 1820 **e.** 2736
6. Bill

Page 11

1. **a.** 209 **b.** 312 **c.** 234 **d.** 1034
2. **a.** $14\times12 = (14\times10)+(14\times2) = 140+28 = 168$ **b.** 299 **c.** 253 **d.** 882
3. **a.** 266 **b.** 624 **c.** 667 **d.** 2048 **e.** 2256 **f.** 3478
4. **a.** 700 cm or 7 m **b.** 700 cm or 7 m **c.** 685 cm or 6 m 85 cm
5. £5·40
6. **a.** 912 km **b.** 1824 km **c.** 1368 km
7. £39
8. **a.** 1200 cm **b.** 12 m
9. 407 cm
10. 255

Page 13

1. **c.** {6, 12, 18, 24, 30, 36, 42, 48, 54, 60, 66, 72, 78, 84, 90, 96}
3. **a.** 16, 22, . . . **b.** 7, 6, . . . **c.** 6, 9, . . . **d.** 8, 4, . . .

Page 15

2. 49 and 81
3. **a.** 36, 16, 100, 121 **b.** 36 **c.** 64

d.

Number	1	3	5	7	8	6	12	11	20
Square	1	9	25	49	64	36	144	121	400

4. $4^2 = 1+2+3+4+3+2+1$
   $5^2 = 1+2+3+4+5+4+3+2+1$
5. **a.** 5 **b.** 6×6 **c.** $7\times7 = 49$ **d.** $8^2 = 8\times8$
6. **a.** $(1+3+5+7)+9 = 16+9 = 25$ **b.** Square
7. **a.** 6 **b.** 2

Page17

1. **a.** 1111 **b.** 2021 **c.** 9909
2. **a.** 27 **b.** 20 **c.** 21 **d.** 2
3. **a.** 500+60+7 **b.** 1000+400+30+2
   c. 5000+60+7 **d.** 2000+40
4. **a.** 342 **b.** 874 **c.** 1346 **d.** 10110 **e.** 4321 **f.** 1203
5. **a.** 99 **b.** 10 **c.** 999 **d.** 100
6. **a.** 1000 **b.** 1000 **c.** 60000 **d.** 61000 **e.** 37000
7. **a.** 100 **b.** 100 **c.** 800 **d.** 900 **e.** 1000 **f.** 11200
8. **a.** 2^4 **b.** 4^2 **c.** 7^4 **d.** 10^5
9. **a.** 25 **b.** 125 **c.** 81 **d.** 32
10. 3^4
11. 7

Page 19

1. **a.** Three **b.** One-nought **c.** One-two **d.** Two-one **e.** One-nought-nought
   f. Two-nought-two
2. **a.** $10_{(5)}$ **b.** $13_{(5)}$ **c.** $20_{(5)}$ **d.** $22_{(5)}$ **e.** $30_{(5)}$
3. **a.** 7 **b.** 12 **c.** 24 **d.** 35 **e.** 56 **f.** 103 **g.** 124
4. 125

5. **a.** 4 friends; 3 marbles left **b.** Yes
6. **a.** $23_{(5)}$ **b.** $31_{(5)}$ **c.** $44_{(5)}$ **d.** $110_{(5)}$ **e.** $111_{(5)}$ **f.** $234_{(5)}$ **g.** $301_{(5)}$ **h.** $1001_{(5)}$ **i.** $2011_{(5)}$
7. **a.** $13_{(5)}$ **b.** $330_{(5)}$ **c.** $302_{(5)}$ **d.** $312_{(5)}$ **e.** $400_{(5)}$ **f.** $310_{(5)}$ **g.** $1010_{(5)}$ **h.** $2000_{(5)}$

Page 21

1. **a.** 25 km, 10 km, 20 km, 15 km
   b. 30 min, 6 min, 24 min

Page 23

1. **a.** 1000 **b.** 500 **c.** 250 **d.** 750
2. **a.** 5 times **b.** 10p
3. **a.** 1000 ml **b.** $\frac{1}{4}$ litre **c.** 100 litres **d.** 14p
4. **a.** 150 litres **b.** £12

Page 25

1. **a.** $\frac{1}{4}$ **b.** $\frac{1}{8}$ **c.** $\frac{1}{2}$ **d.** 20 **e.** 5 **f.** 10
2.

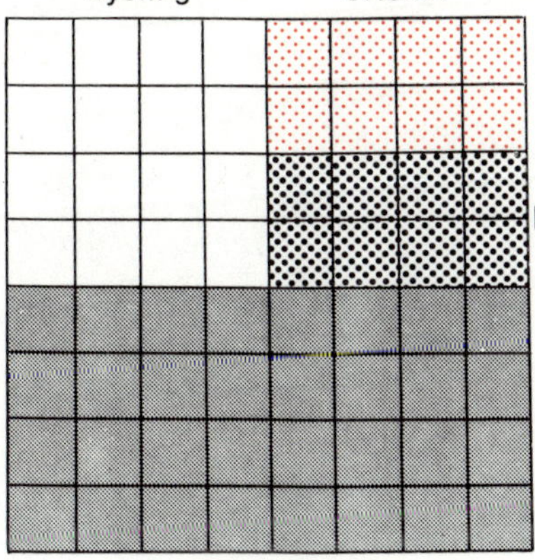

3. **a.** $\frac{3}{4}$, 24 matches **b.** $\frac{1}{8}$, 4 matches **c.** $\frac{1}{8}$, 4 matches
   d.

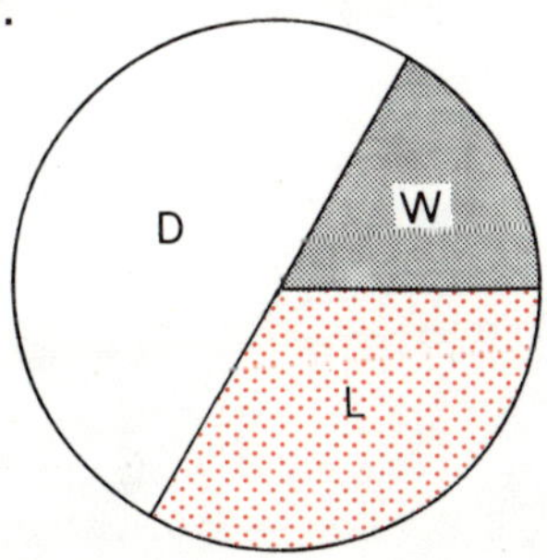

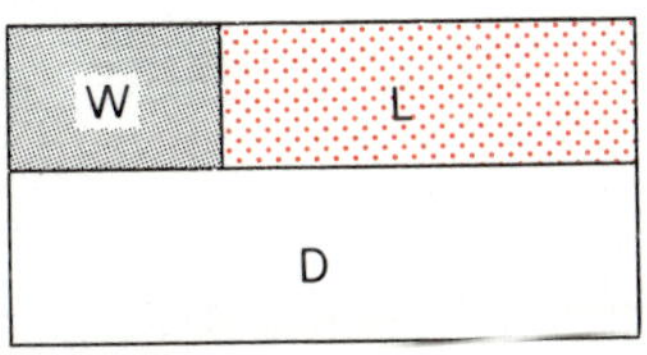

4. **a.** $\frac{1}{4}$ **b.** $\frac{1}{8}$ **c.** $\frac{1}{8}$ **d.** $\frac{1}{8}$ **e.** $\frac{3}{8}$
   f. Total area: 128 m²
   Playroom: 16 m²
   Dining-room: 16 m²
   Hall: 16 m²
   Kitchen: 32 m²
   Lounge: 48 m²

   g.

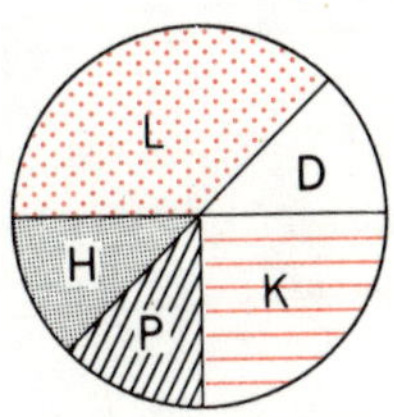

Page 27

1. a. 12 b. 12 c. 2×5 d. 4×6 = 24
2. a. 10 b. 8 c. 30 d. 50 units
3. a. 20 cm b. 15 cm c. 300 cm²
4. a. 6 cm × 2 cm = 12 cm² b. 6 cm²

Page 29

1. a. Unit, 5 b. $\frac{1}{10}$th, $\frac{5}{10}$ c. $\frac{1}{100}$th, $\frac{5}{100}$
   d. Unit, 5 e. $\frac{1}{10}$th, $\frac{5}{10}$ f. $\frac{1}{100}$th, $\frac{5}{100}$
2. a. $<$ b. $=$ c. $<$ d. $<$
3. a. 39·81 b. 44·22 c. 156·75 d. 319·77 e. 1099·89
4. a. 50·93 b. 59·87 c. 59·07 d. 300·75 e. 788·81
5. a. 12·5 b. 125 c. 1250 d. 0·125 e. 4·702 f. 0·4702 g. 470·2
   h. 4702
6. a. 1·2, 1·5 b. 0·4, 0·45 c. 0·23, 0·27
7. a. 14·53 b. 10·5 c. 120·51 d. 60·5
8. a. £2 b. £4
9. a. 25p b. £2·50
10. 2·05 litres

Page 31

1. a. £2·40 b. £8·75 c. £16·56 d. £0·40$\frac{1}{2}$ e. £20·50
2. a. £1·05 b. £0·22 c. £0·11 d. £1·15
3. a. £7 b. £2·40 c. £0·35 d. £3·50

	e. £	f. £
	0·70	7·00
	1·05	10·50
	0·48	4·80
	2·23	22·30

4. a. £3·96 b. £9·90 c. £3·24
5.

a. £	b. £	c. £
0·40	2	0·48
0·35	1	6·00
2·00	6	0·35
2·75	9	6·83

Page 33

1. a. 3 b. 1 c. 16 d. 1 e. 7 f. 72
2. a. $>$ b. $<$ c. $=$ d. $<$ e. $<$ f. $=$
3. a. − b. × c. ÷ d. × e. − or ÷ f. + or − or ×
4. Closed: a, c, d, e, f
   False: e, f
5. a. $\triangle$10 +5 = 15 b. $\triangle$30 −10 = 20 c. $\frac{\triangle 20}{2} = 10$
6. c. B

Page 35

2. a. 1·2 m b. 1·02 m c. 2·25 m d. 2·52 m
3. a. 1 m 25 cm b. 2 m 56 cm c. 2 m 65 cm d. 3 m 5 cm
5. a. 8 m 35 cm b. 79 cm c. 11 m 50 cm

Page 37

3.

Number	9	10	11	12	13	14	15	16	17	18	19	20
Square	√	×	×	×	×	×	×	√	×	×	×	×
Rectangular	√	√	×	√	×	√	√	√	×	√	×	√
Different arrays	1	2	0	4	0	2	2	3	0	4	0	4

4. {4, 9, 25, 49}
5. Not true; 81 is 3×27
6. a. P = {4, 9, 16, 25}
   b. Q = {4, 6, 8, 9, 10, 12, 14, 15, 16, 18, 20, 21, 22, 24, 25, 26, 27, 28}; Yes
   c. R = {9, 25} d. S = {4, 16}
7. a. 2 b. They are even numbers.

Page 39

4. b. Yes; Yes
6. b. Yes c. Yes
7. a. 3 b. 2 c. 6 d. Trapezium, Rhombus, Equilateral triangle
8. A square

Page 41

1. a. B and E b. A, C, D, and E
5. a. Yes b. Only the 6-sided figure will tessellate.

Page 43

1. a. 9 b. 10 c. 14
2. a. $\{\frac{1}{5}, \frac{2}{10}, \frac{2}{7}\}$ b. $\{\frac{3}{2}\}$ c. $\{\frac{1}{5}, \frac{2}{10}, \frac{2}{7}\}$ d. $\{1\frac{1}{2}, 3\frac{1}{2}\}$ e. $\{1\frac{1}{2}, \frac{3}{2}, 3\frac{1}{2}\}$ f. { }
3. a. $3\frac{1}{4}$ b. $7\frac{1}{5}$ c. $6\frac{2}{7}$ d. $8\frac{8}{9}$ e. $30\frac{1}{7}$ f. $31\frac{1}{10}$ g. $40\frac{5}{6}$ h. $72\frac{1}{8}$
4. a. $1\frac{7}{10}$ b. $2\frac{3}{5}$ c. $1\frac{3}{100}$ d. $3\frac{7}{20}$ e. $2\frac{14}{15}$
5. a. $3\frac{1}{2}$ m b. $2\frac{1}{4}$ h

Page 45

1. a. 5 b. 25 c. 20 d. 7
2. a. 7 b. 5 c. 4 d. 5
3. a. 3 b. 7 c. 75
4. a. $A \cap B = \{1, 2, 4\}$ b. $\frac{2}{5}$
5. a. $A = \{1, 2, 3, 5, 6, 10, 15, 30\}$
   $B = \{1, 2, 4, 5, 10, 20, 25, 50, 100\}$
   $A \cap B = \{1, 2, 5, 10\}$
   b. $\frac{3}{10}$
6. a. $\frac{1}{2}$ b. $\frac{1}{2}$ c. $\frac{3}{10}$ d. $\frac{3}{25}$ e. $\frac{7}{50}$
7. a. $3\frac{1}{2}$ b. $1\frac{1}{10}$ c. $2\frac{1}{5}$ d. $1\frac{3}{5}$ e. $5\frac{3}{10}$
8. Yes
9. No
10. $1\frac{3}{4}$ litre

Page 47

1. a. (1, 2) b. Eye c. No d. (1, 1) e. Tom's head would fall off.
2. a. MATHSISFUN
3. a. {(2, 2), (3, 1), (3, 3), (4, 2)} b. A relation d. (3, 2) e. Four lines
4. a. (5, 3) b. (1, 4) c. (3, 6)

Page 49

1. a.

×	2	4	6	8
7	14	28	42	56

b.

×	1	3	5	7
9	9	27	45	63

2. a. 6 rem. 2 b. 5 c. 4 rem. 31 d. 7 e. 7 rem. 4
3. a. = b. > c. < d. > e. < f. >
g. a. 10 b. 10 rem. 1 c. 9 rem. 24 d. 10 rem. 22 e. 9 rem. 20 f. 14 rem. 6
4. 15 litres
5. a. 10 b. 10 c. 1000
6. 45p
7. 40
8. a. 24 b. 14 c. 28
9. 155
10. 74

Page 51

1. a. 24p b. 8p c. {(1, 8), (2, 16), (3, 24)} d. 80p f. 48p, 72p, 64p
   g. 3, 5, 4, 7
2. a. {(0, 0), (4, 30), (8, 60), (12, 90)} c. 120 min, 75 min, 15 min, 45 min, 105 min
   d. 6 km, 2 km, 12 km, 10 km, 1 km
3. 18, 9, 3, 6

Page 53

1. a. 2×6, 3×4, 4×3, 6×2 b. 2×8, 4×4, 8×2
2. a. R = {2, 3, 5, 7, 11, 13, 17, 19, 23}
   b. S = {2, 3, 5, 7, 11, 13, 17, 19, 23, 29, 31, 37, 41, 43, 47}
   c. n(R) = 9 d. n(S) = 15
3. a. 53, 113 b. 97, 139
4. a. 2, 3 b. 3 c. 2, 5 d. 2, 3 e. 2, 7 f. 3, 5
5. a. 6 b. 15 c. 10 d. 4 e. 12 f. 18
6. Yes; No
7. a. {2, 3} b. {2} c. {2, 3}
8. a. Yes b. Prime

Page 55

1. 100 cm, 55 cm, 180 cm, 125 cm
2. a. 6 m b. 3 m c. 60 m d. 600 m e. 10 f. 10 g. About 11 h. 100
3. a. 45 m b. 55 m c. 400 paces, 200 m d. About 22 m

Page 57

1. a. $\frac{9}{10}$ b. $\frac{3}{10}$ c. $\frac{12}{10} = 1\frac{\cancel{2}^{1}}{\cancel{10}_{5}}$ d. $\frac{\cancel{2}^{1}}{\cancel{10}_{5}}$ e. $\frac{\cancel{8}^{4}}{\cancel{10}_{5}}$
2. a. 10 b. 10 c. 10 d. 20 e. 100
3. a. $\frac{11}{20}$ b. $\frac{9}{10}$ c. $\frac{\cancel{15}^{3}}{\cancel{20}_{4}}$ d. $\frac{41}{100}$ e. $\frac{\cancel{45}^{9}}{\cancel{100}_{20}}$
4. a. $\frac{1}{10}$ b. $\frac{1}{20}$ c. $\frac{\cancel{5}^{1}}{\cancel{20}_{4}}$ d. $\frac{17}{100}$ e. $\frac{3}{20}$
5. $\frac{\cancel{8}^{4}}{\cancel{10}_{5}}$, $\frac{3}{10}$, $\frac{1}{2}$
6. a. $\frac{21}{20}$ m = $1\frac{1}{20}$ m b. $\frac{9}{20}$ m
7. a. $1\frac{1}{2}$ kg b. Jane's is $\frac{1}{10}$ kg heavier.

Page 59

2. a. $\frac{9}{10}$ b. $\frac{4}{5}$ c. $\frac{7}{10}$ d. $\frac{3}{5}$ e. $\frac{4}{5}$ f. $\frac{9}{10}$
3. a. $2\frac{1}{4}$ b. $1\frac{4}{5}$ c. $4\frac{2}{3}$ d. $2\frac{4}{5}$ e. $4\frac{1}{2}$ f. $3\frac{1}{10}$
4. a. $\frac{1}{4}$ b. $\frac{2}{5}$ c. $\frac{9}{10}$ d. $\frac{7}{4}$
5. a. 25p b. 75p
6. $2\frac{1}{4}$ h
7. $\frac{6}{7}$ week or 6 days
8. a. 200 b. 300
9. a. 500 b. 200 c. 800 d. 12

Page 61

1. a. 4, 1, 4, 0, 2, 0 b. {A} c. {C} d. {A, B, C, D, E, F} e. { }
2. Four
3. Four
6. a. 6 b. 6

Page 63

1. a. 2500 kg b. 1400 kg c. 1450 kg d. 500 kg
2. a. 2 t b. 2·5 t c. 1·3 t d. 0·6 t
3. 61 500 kg
5. a. 1000 g, 1 kg b. 5000 g, 5 kg c. 4 kg d. $\frac{4}{5}$, $\frac{1}{5}$

Page 65

1. a. 0·09 b. 0·16 c. 0·75 d. 1·9 e. 1·75 f. 1·25
2. a. 14 b. 4 c. 7
3. a. $\frac{7}{20}$ b. $10\frac{3}{50}$ c. $1\frac{3}{50}$ d. $2\frac{9}{20}$ e. $\frac{13}{20}$ f. $10\frac{13}{100}$
4. a. 0·22 b. 0·15 c. 0·16 d. 0·85 e. 0·98 f. 0·96
5. a. $\frac{27}{100}$ m b. 27 cm
6. 25 cm = $\frac{25}{100}$ m = 0·25 m
7. a. 0·25 = $\frac{1}{4}$ b. 0·45 = $\frac{9}{20}$ c. 1·3 = $1\frac{3}{10}$ d. 2·24 = $2\frac{6}{25}$
8. a. 0·6 = $\frac{3}{5}$ b. 0·75 = $\frac{3}{4}$ c. 0·25 = $\frac{1}{4}$ d. 2·4 = $2\frac{2}{5}$
9. a. $\frac{7}{10}$ b. $\frac{3}{10}$ c. 0·7 d. 0·3

Page 67

2. ∠POC or ∠COP; ∠MQN or ∠NQM; ∠RAT or ∠TAR; ∠XZY or ∠YZX
3. a. {r} b. {p, q} c. {s} d. {q, r, s}
4. A – C, B – F, E – D
8. Three

Page 69

1. a. Base five: 20, 21, 41, 110, 113, 124, 131, 144
   b. Base ten: 10, 11, 15, 25, 27, 35, 45, 49
2. a.

+	0	1	2	3	4
0	0	1	2	3	4
1	1	2	3	4	10
2	2	3	4	10	11
3	3	4	10	11	12
4	4	10	11	12	13

b.

−	0	1	2	3	4
4	4	3	2	1	0
10	10	4	3	2	1
11	11	10	4	3	2
12	12	11	10	4	3
13	13	12	11	10	4

3. a. $41_{(5)}$ b. $203_{(5)}$ c. $420_{(5)}$ d. $1014_{(5)}$ e. $1000_{(5)}$
4. a. $33_{(5)}$ b. $14_{(5)}$ c. $41_{(5)}$ d. $34_{(5)}$ e. $202_{(5)}$
5. Base five: a. 42 b. 133 c. 304 d. 441 e. 440 f. 1300 g. 2022
6.

4	14	2
3	10	12
13	1	11

11	1	13
12	10	3
2	14	4

Page 71

1. a. 45 min
   b. 1 h 25 min
   c. 2 h 10 min
   d.

AXTON	10.15	13.10
BELTON	10.55	13.50
CETON	11.40	14.35
DETON	12.25	15.20

   e. 1.10 p.m., 1.50 p.m., 2.35 p.m., 3.20 p.m.
2. a. 7 h 30 min b. 1 h 44 min c. 12 h 30 min d. 15 min

Page 73
1. $\overline{AB}$ and $\overline{EF}$, $\overline{CD}$ and $\overline{JH}$, $\overline{OP}$ and $\overline{IG}$, $\overline{NM}$ and $\overline{KL}$
2. A, B, C
3. A parallelogram
5. **a.** They are the same size. **b.** They are the same size. **c.** Two **d.** Four

Page 75

1.

a.	8	6	2	√	
b.	6	4	2	√	
c.	4	6	2		√
d.	11	15	4		√
e.	10	12	2		√
f.	9	5	4	√	

2. **a.** $^{-}2$ **b.** $^{+}2$ **c.** $^{-}5$ **d.** $^{+}5$ **e.** $^{+}7$ **f.** 0 **g.** (4, 2), (4, 2), (3, 6), (3, 7), (6, 8)
3. **a.** (6, 3) **b.** (3, 6) **c.** (6, 2) **d.** (9, 3)
4. **a.** (4, 6) **b.** (2, 7)
5. Any pair (n, n+2)

Page 77
1. **a.** 1 **b.** 1 **c.** 0 **d.** 1 **e.** 1 **f.** 4 **g.** 1 **h.** 2
3. No
4. 1
5. 0
6. **a.** C **b.** B, D
7. **a.** **b.** **c.** **d.**

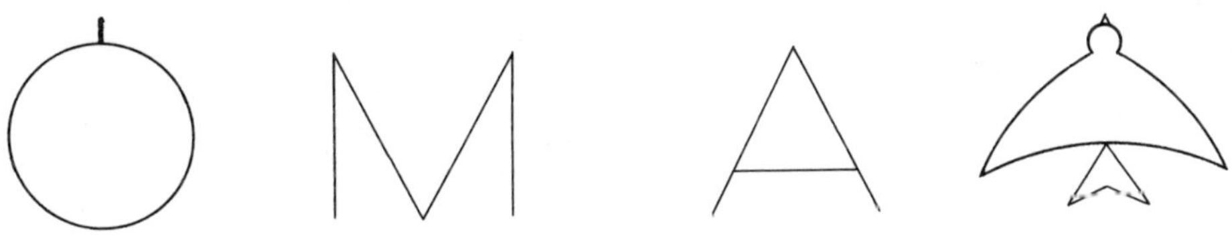

Page 79
1. **a.** 0 **b.** 3 **c.** 5 **d.** 0 **e.** 6
2. **a.** 2 **b.** 6 **c.** 5 **d.** 1 **e.** 3 **f.** 2 **g.** 0 **h.** 0 **i.** 0
3. **b.** Channel 2 **d.** 5 o'clock

Page 81
3. **a.** Yes **b.** Yes
4. **a.** Six **b.** Yes
5. **a.** C has 2, D has 1 **b.** C
6. **a.** E and H **b.** F and G **c.** F and G
7. **a.** One **b.** One **c.** As many as you like
8. Shape L

Page 83

1. **a.** 32 **b.** 18 **c.** 21
   16 9 10
   16 9 11
2. 7, 6, 12, 8, 8 cm^2
3. 15, 7, 13, 16, 16 cm^2
4. **a.** 16 m^2 **b.** 8 m^2 **c.** 8 m^2

Page 85

2. **a.** 12 **b.** 5 **c.** 10 **d.** 12 **e.** 12 **f.** 8
3. **a.** 2 **b.** 2 **c.** 1 **d.** 1

Page 87

1. **a.** 4 **b.** 3 **c.** 4 **d.** 0 **e.** 4
2.

+	0	1	2	3	4
0	0	1	2	3	4
1	1	2	3	4	0
2	2	3	4	0	1
3	3	4	0	1	2
4	4	0	1	2	3

x	0	1	2	3	4
0	0	0	0	0	0
1	0	1	2	3	4
2	0	2	4	1	3
3	0	3	1	4	2
4	0	4	3	2	1

4. **a.**

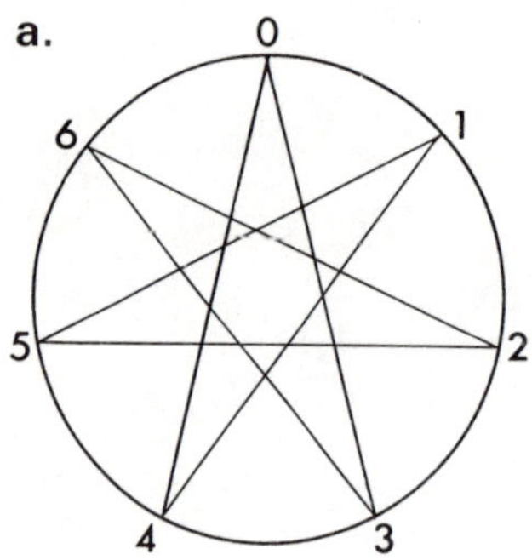

b.

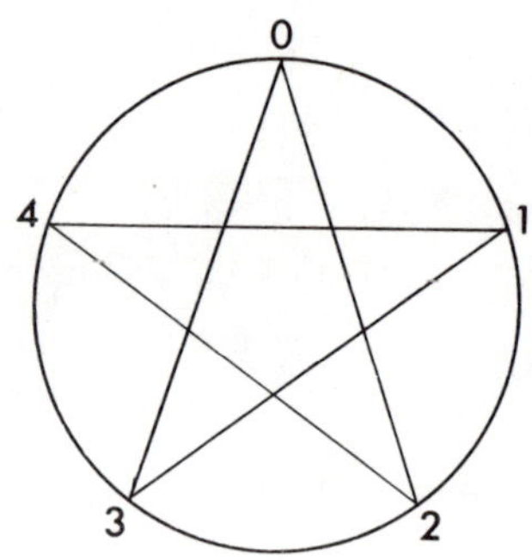

5. **a.**

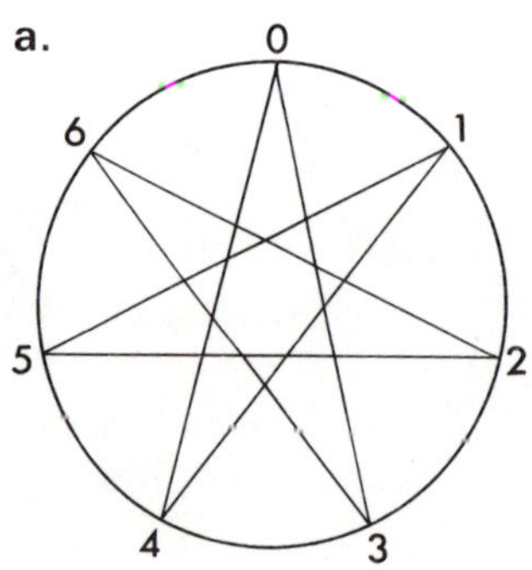

b.

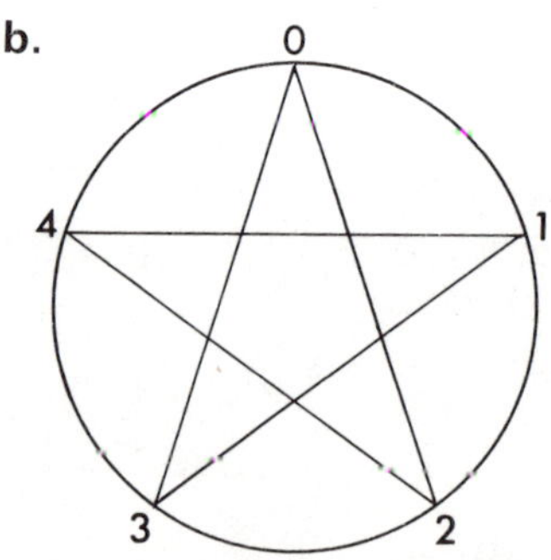

6. **4a** and **5a** are similar. **4b** and **5b** are similar.

Page 89

1. a. 20 b. 20 c. 3 d. 60 e. 60 cm³
2. a. 40 b. 200 c. 160 d. 4, 20, 16 packets e. 20p, £1, 80p
3. a. 1000 b. 125 c. 1000 cm², 125 cm²

Page 91

1. 50, 50
2. a. 72 g b. 36 g c. 144 g d. 50·4 g
3. a. 20p, 72 g b. 80p, 288 g c. 72 g d. 288 g
4. a. £1 b. 10p c. 50p d. £10
5. d. weigh, area e. Four
6. a. 2p, 1p, 1p b. 10 g

Page 93

1. a. 50 b. 50 c. 20 d. $12\frac{1}{2}$ e. 10 f. 20 g. 75p h. 150p
2. a. £1·50 b. £2·50 c. £9·50 d. £12·50 e. £12 f. £0·75 g. £1·25
3. a. £17 b. £31 c. £2·50 d. £225 e. £14 f. £4·50 g. £12·50
4. a. 63p b. £1·52 c. £1·$72\frac{1}{2}$ d. £1·17 e. £2·$36\frac{1}{2}$ f. £6·93

5\.

	a.		b.
	p		p
	10		95
	50		86
	36		$37\frac{1}{2}$
	31		42
	£1·27		£2·$60\frac{1}{2}$

6. a. 14p b. 11p
7. a. £612·50 b. £3 c. £1·60 d. 48p e. 56p f. 48p

Page 95

3. a. D b. B c. C d. Four e. Eight
4. Circle
5. No
7. Centre of the sphere

Page 97

1. a. 400 b. 320 c. 200 d. 600
2. a. 310 g b. 875 g c. 625 g d. 440 g
3. a. About 31 cm², about 310 cm² b. Both weigh about 100 g
4. a. 2 cm b. $1\frac{1}{2}$, 2, 3 kg c. 1500, 2000, 3000 g d. 1 kg e. Nearly 3 months
   f. Just over 4 months

Page 99

1. Real lengths: a. 5 m b. 3 m c. 3·5 m d. 2·5 m
3. AB = 5 km, CD = 7 km, EF = $4\frac{1}{2}$ km, GH = $4\frac{1}{2}$ km
4. a. m($\overline{AB}$) = 3 cm, m($\overline{BC}$) = 6 cm b. 3 km, 6 km c. E d. E, SW

Page 101

2. **a.** 3 **b. and c.** A line segment **d.** Yes
3. **a.** BCGF **b.** ABFE and ABCD **c.** $\overline{BC}$ **d.** 3 **e.** 9 **f.** 7
4. **a.** 36 cm **b.** 54 cm^2

Page 103

2. **a.** 1% **b.** 10% **c.** 5% **d.** 50% **e.** 15% **f.** 100%
3. **a.** $\frac{1}{5}$, 0·2 **b.** $\frac{2}{5}$, 0·4 **c.** $\frac{1}{4}$, 0·25 **d.** $\frac{3}{4}$, 0·75
4. **a.** 5% **b.** 15% **c.** 80% **d.** 75%
5. **a.** $\frac{1}{2}$, 50% **b.** $\frac{1}{4}$, 25%

Page 105

1. **c. and d.** {$^-15$, 0, $^+15$}
2. **a.** $^+11$ **b.** $^-3$ **c.** $^+7$ **d.** $^+7$ **e.** $^+11$ **f.** $^-5$
3. **a.** (i) No, (ii) Yes **b.** $^+4$ °C

c.

$^+5$	$^+5$	$^+5$	$^+2$	$^-5$
$^+5$	$^-5$	$^-10$	$^-5$	$^+10$
$^+10$	0	$^-5$	$^-3$	$^+5$

Page 107

1. **a.** 4·2 cm, 3 cm, 3 cm; 4·2 km, 3 km, 3 km **b.** E, W, NE
2. **c.** Two
3. **a.** 5 km

Page 109

1.

A	B	C	D	E
Yes	Yes	No	Yes	Yes

2. **a.** J **b.** F, I, K

c.

Network	F	G	H	I	J	K
V	4	3	1	5	3	12
E	4	4	2	5	3	12
R	2	3	3	2	2	2

d. V−E+R = 2